Environmental Science: Active Learning Laboratories and Applied Problem Sets

Travis Wagner, Ph.D.
and Robert Sanford, Ph.D.

Department of Environmental Science
University of Southern Maine

WILEY

JOHN WILEY & SONS, INC.

"Those who dwell, as scientists or laymen,
among the beauties and mysteries of
the earth are never alone or
weary of life."

from *The Sense of Wonder* by
Rachel Carson (1907–1964).

To order books or for customer service call 1-800-CALL-WILEY (225-5945).

ISBN 0-471-67191-6

10 9 8 7 6 5 4 3

Acknowledgements

Special thanks are due to our colleagues, Dr. Samantha Langley-Turnbaugh (Chairperson, Department of Environmental Science, University of Southern Maine [USM]), and Dr. Christopher Pennuto (SUNY College at Buffalo), who were instrumental in developing and field-testing a number of the labs. We appreciate the patience of our students as we tried out various versions of these laboratory and homework activities. Dr. Maureen Malachowski (consulting industrial hygienist) and our departmental colleagues Dr. Sharon D'Orsie and Dr. Merrie Cartwright provided ideas, comments, and suggestions. Eileen Burk, USM undergraduate research assistant, provided research support. Christine Sears, USM art student, provided graphic design. Dr. Dick Stebbins and the Maine Mathematics—Science Teaching Excellence Collaborative—inspired us to re-think our learning objectives and strategies. Thanks to Dr. Jane Heinze-Fry (Scholar in Residence, Emerson College) for advice on concept maps and to Dr. Grace Eason (University of Maine, Farmington) for her suggestions. Librarians Robin Sanford (University of New England) and Tim Lynch (University of Southern Maine) helped track down references. Helpful comments came from the reviewers provided by John Wiley & Sons. We appreciate the encouragement and support of Geraldine Osnato, Projects Editor at John Wiley & Sons. Thanks to Paul F. Miller. Thanks to SENCER (Science Education for New Civic Engagements and Responsibilities), an NSF-funded initiative of the American Association of Colleges and Universities, for its inspirational efforts.

Cover photo by Libby Barrett

Preface

Environmental Science: Active Learning Laboratories and Applied Problem Sets is designed to introduce environmental science students to the broad, interdisciplinary field of environmental science by presenting specific labs that use natural and social science concepts to varying degrees and by encouraging a "hands on" approach to understanding the impacts from the environmental/human interface. The laboratory and homework activities are designed to be low-cost and to reflect a sustainability approach in practice and in theory.

In addition to the overall approach and design, this text:

■ Utilizes innovative learning techniques, such as problem-based, active, and critical learning. Group and cohort paths to knowledge are encouraged. As part of this approach, we stress student-initiated inquiry and experimentation as well as emphasizing civic responsibility in environmental science.

■ Develops a variety of topics that mirrors the variety of subjects found in environmental science, including urban ecology, global impacts, air pollution, solid waste, energy consumption, soils identification, water quality assessment, and the scientific method.

■ Encourages students to grasp the big picture by relating the lab activity to real life conditions and their individual contribution to environmental problems. We have individual measures and descriptions, but we also nurture application of this learning to the larger ecological picture.

■ Develops a variety of techniques that include traditional laboratory activities, field exercises, Internet research, calculations/extrapolations, and critical analysis. Because the pursuit of real-world environmental science involves all these components, so do our lab activities.

■ Emphasizes the improvement of written and other forms of communication. So much of science has become participatory, particularly in making decisions about its application (i.e., environmental policy). We provide ways for the student to discover that the communication of scientific information is as important as the acquisition of scientific knowledge.

■ Contains relevant problem sets that can be used as labs, lab supplements, or homework assignments for environmental science lectures.

Table of Contents

Introduction

Environmental science is the study of how humans and nonhumans interact with one another and with the nonliving environment. An important element of environmental science is that it views this relationship and interaction as interconnected—a gigantic system. Based on this perspective, environmental science is an interdisciplinary science, which integrates concepts from other disciplines, including biology, chemistry, ecology, economics, engineering, ethics, geology, physics, policy science, sociology, and toxicology. Environmental science is an experimental science, meaning that its study is based on the application of the scientific method.

Thus, the labs and problem sets contained in this manual are based on the following:

- Students need to understand basic concepts associated with the interdisciplinary and interconnected nature of environmental science.

- To obtain the necessary evidence for environmental problems, students must understand and experience the scientific method.

- Students need to recognize that social science plays a major role in environmental science because environmental problems are socially constructed.

- Individuals do matter. We all have personal impacts on the environment; collectively recognizing personal responsibility is crucial for change.

The labs and problem sets contained in the manual represent the broad spectrum of the interdisciplinary and interconnected nature of environmental science. The subject matter of the labs and problem sets is based on Botkin and Keller's (2003) claim that certain issues are basic to the study of environmental science. These include:

- Rapid human population growth, in conjunction with affluence and technology, is a fundamental cause of environmental problems.

- Human beings affect the environment of the entire planet; therefore a global perspective on environmental problems must be taken.

- Urban environmental issues and their effects need to be given primary focus.

- Sustaining our environmental resources is crucial for future availability.

- Policy solutions to environmental problems require making value judgments based on knowledge of scientific facts.

These laboratory exercises and problem sets have been designed to use a minimum of equipment at a minimum of cost. This was out of necessity from our own experiences and is in keeping with our views of sustainability and wise resource use. The laboratory and problem sets can be used in a variety of ways; some problem sets may be done as laboratory activities and vice versa. Some problem sets support particular laboratory activities, but there is a great deal of flexibility in how this is done, and we have merely provided general topical subheadings in the Table of Contents. Every environmental science instructor has his or her own "tool-kit" of favored laboratory activities; some are so essential to understanding environmental science that they have become "classics." We have made every attempt to provide original exercises and problem sets while still paying homage to the time-honored classics that appear to belong to the collective whole.

The Internet is used here as a portal to information, but with the appropriate caveats that as a portal it does not discriminate between good or junk evidence. Hence, the student must become a careful consumer of information and practice *caveat emptor*. And in fact, that is a key task—we are awash in a sea of environmental information, and we need to sort our way through it. Consequently, we try to use only those Internet Web sites that appear stable and reliable and that serve as springboards to further information.

Reference

Botkin, D.B. and E.A. Keller. 2003. Environmental Science: Earth as a Living Planet. 4th ed. John Wiley & Sons, New York.

General Information

Laboratory Health & Safety Procedures

Every laboratory has its own set of rules. Many rules are similar to the following:

1. Never smell, taste, or touch an unknown substance.

2. Do not bring food or beverages to the laboratory.

3. Clean up your work area when the lab period is complete.

4. Wash your hands before leaving the laboratory.

5. Keep lab benches free of extraneous books and clothing.

6. Do not pipette by mouth at any time.

7. When using microscopes show instructor its condition before leaving. Only use lens paper for cleaning lenses.

8. Know location of first-aid kit and how to use it.

9. Know location of fire extinguisher and/or fire blanket and how to use them.

10. Know location of eyewash and/or emergency shower stations and how to use them.

11. Contact lenses are a concern depending on the chemicals used or stored in the lab. Be sure to check with the instructor on the use of contact lenses at the beginning of the course.

12. Do not apply make-up or other products to your skin in the lab.

13. Wear appropriate protective clothing; do not wear open-toed shoes or sandals.

14. Dispose of materials in proper containers: paper in recycling bins, glass in the special "broken glass" bin, and biohazards in the "biohazard" bin.

15. Report all chemical spills to the instructor. Spills will be handled by the instructor in accordance with university procedures.

16. Take care and remain attentive in the lab at all times.

WRITING LABORATORY REPORTS

Writing is one of the most important things you will do in this class. Generally, there are two types of write-ups: formal laboratory reports and informal laboratory write-ups. All written lab-related materials must be typed and submitted to the instructor in hard copy. Unless instructed otherwise, assume no electronic submissions will be accepted.

FORMAL LABORATORY REPORTS

Laboratory reports are your tool for expressing what you did, why you did it, and what you learned in the process. Even if your understanding of the procedure, techniques, and results is perfect and your results are error-free, a poorly written report will suggest that you did not understand what you have done. Good writing is good writing, be it creative fiction, an editorial, journal article, or scientific communication. Writing reports is not difficult if you remember a few guidelines about writing and the structure of a good report.

Your formal lab report <u>must</u> have the following components in the following order.

1. Title

2. Introduction

3. Materials and Methods

4. Results

5. Discussion

6. Literature Cited (or References)

   Note: Metric measurements are used in science for presenting facts and figures. When presenting information to the general public (i.e., not in a formal lab report or scientific writing), English units are often used.

> An excellent resource to help you prepare, organize, interpret, and write your formal laboratory report is Labwrite. Labwrite is an instructional project originating from North Carolina State University and sponsored by the National Science Foundation. The Labwrite Site is: http://labwrite.ncsu.edu/www/

TITLE

What Did You Study?

The title of a lab report should indicate exactly what you studied:

For example, The Effects of Dissolved Oxygen on Brown Trout (*Salmo trutta*) Survival. This title explains the environmental factors manipulated (dissolved oxygen), the parameter measured (survival), and the specific organism used (*S. trutta*).

> *Note:* Always use the scientific (Latin) name for plants, animals, and other biota. Scientific names must be in *italics* or <u>underlined</u>. The genus is capitalized and the species is in lower case. After the first use, scientific names can be shortened by using the first letter of the genus followed by the complete species name. For example, the scientific name for brook trout is *Salmo trutta* (or <u>Salmo</u> <u>trutta</u>). When referring to the brook trout again in a lab report, use *S. trutta* (or <u>S. trutta</u>) or the common name.

INTRODUCTION

Why Did You Study This Phenomenon?

The introduction should identify the phenomenon you studied/tested and provide relevant background information (e.g., why did you study it, what is its environmental relevance?). This is likely to include information from other studies or documents, which <u>must</u> be properly referenced. The introduction should end with your hypothesis, which is a sentence that specifically and clearly states the question your experiment was designed to answer.

For example:

> Our hypothesis was that precipitation with a pH of 3.6 would significantly reduce the success of red maple (*Acer rubrum*) seed germination.

MATERIALS AND METHODS

What Did You Do? How Did You Do It?

In the materials and methods section of a formal lab report, you will describe how and when you did your work, including experimental design, experimental apparatus/equipment, methods of gathering and analyzing data, and types of experimental control. This section must include complete details and be written clearly enough to allow readers to <u>duplicate</u> the experiment; thus, it must be written in chronological order. This section should be written in the past tense because you have already done the experiment. It should <u>not</u> be written in the form of instructions or as a list of materials as in a laboratory manual. Instead, it is written as a narrative, which describes in the active voice what you did. For example, (active voice) *We filled six petri dishes with 20 ml of tap water in each.* And be sure to use metric measurements.

RESULTS

What Did You Find?

In the results section, you present your observations and data with <u>no</u> interpretations or conclusions about what they mean. Tables and graphs should be used to supplement the text and to present the data in a synthesized, more understandable form. Use the past tense to describe your results.

When using tables or figures, each table or figure <u>must</u> be introduced within the text. All tables and figures must be numbered and have self-explanatory titles so that the reader can understand their content without the text (e.g., Table 1. Percent of red maple [*Acer rubrum*] seedlings exhibiting visible injury after exposure to water with a pH of 3.6). Assign numbers to tables and figures in the order they are mentioned in the text. Tables and figures are numbered independently of each other (i.e., Table 1 and 2, and then Figure 1 and 2). Tables are labeled at the top and figures at the bottom. Tables are referred to as tables; all other items (graphs, photographs, drawings, diagrams, maps, etc.) are referred to as figures.

DISCUSSION

What Does It Mean?

Explain what you think your data mean. Describe patterns and relationships that emerged. It is also very important to explain how any changes to or problems with the experimental design/procedure may have affected the results.

REFERENCES OR LITERATURE CITED

Use the format below to cite any literature used in your report (e.g., your textbook, journal articles, books, and so forth). In the text of your report, if you cite specific information, or quote data or persons, cite references using the author's surname, and year of publication (e.g., Botkin and Keller, 2003); give the page number for the source quoted or paraphrased (to help avoid plagiarism—for example, Botkin and Keller, 2003, p. 234). Sources must be credited if you obtain ideas or thoughts from them, even if you are not giving a direct quote.

No single, correct citation format exists; each discipline (e.g., biology, economics, etc.) and each journal has a preferred citation format. Thus, what is presented below is one particular style, but you will see other styles in your research. The most important aspect is to be consistent and to ensure all necessary information is contained in the citation.

For the purposes of this textbook, when you write lab reports use the following citation format from the Soil Science Society of America Journal (it is a typical science format[1]), unless your instructor recommends a different one.

Journal Articles:
Single Author
Connell, J.L. 1974. Species Diversity in Tropical Coral Reefs. Science 234:23–26.

Multiple Authors
O'Rourke, D., L. Connelly, and C.P. Koshland. 1996. Industrial Ecology: A Critical Review. International Journal of Environment Pollution 6:89–112.

Books:
Single Author
McHarg, I.L. 1971. Design with Nature. Doubleday, Garden City, NY.

Multiple Authors
Botkin, D.B. and E.A. Keller. 2003. Environmental Science: Earth as a Living Planet. 4th ed. John Wiley & Sons, New York.

American Society of Agronomy, Crop Science Society of America, Soil Science Society of America. 1998. Publications Handbook & Style Manual. ASA, CSSA, SSCA, Madison, WI.

Chapter in a Book
Rabe, G.B. 1999. Sustainability in a Regional Context: The Case of the Great Lakes Basin. p. 248–281. In D.A. Mazmanian and M.E. Kraft (ed.) Towards Sustainable Communities: Transition and Transformations in Environmental Policy. MIT Press, Cambridge, MA.

Internet:
Citations for Internet (web) sites should be similar to print media citations, including author, publication date, article title, site title, URL, and date the information was posted (or when the address was accessed).

Internet Article:
Sanchirico, J.N. and R.G. Newell. 2003. Catching Market Efficiencies: Quota-Based Fisheries Management. Resources 150 [Online], 23 Sept. 2003. Available at http://www.rff.org/rff/Documents/RFF-Resources-150-catchmarket.pdf. (verified 2 June 2004).

---

[1]Source: American Society of Agronomy, Crop Science Society of America, Soil Science Society of America. 1998. Publications Handbook & Style Manual. ASA, CSSA, SSCA, Madison, WI.

Government Website:
U.S. Environmental Protection Agency (US EPA). 2004. Home Page [Online]. Available at http://www.epa.gov (verified 3 June 2004).

INFORMAL LABORATORY WRITE-UPS

Some of the laboratory exercises are not formal experiments and thus are not written up as a formal laboratory report. These "informal" lab exercises contain a series of questions designed to promote a particular experience or a range of exploration for an environmental issue. Some labs require you to respond merely to the questions, whereas other labs require you to use the questions to guide the development of your response in narrative form or a memorandum. Be sure your answers are complete and, when appropriate, show your work. When answering questions, be sure to write out full answers. Never simply state *no* or *yes*, but define yourself or explain why a No or Yes response is appropriate. Provide references for any information that comes from other sources.

PART TWO

Labs

CHAPTER
ONE

Environmental Awareness

OBJECTIVES

- Be able to observe and describe basic environmental conditions in a local setting
- Be able to locate environmental information in the library or on the Internet

INTRODUCTION

A society that knows its natural environment is better able to understand and address environmental problems. This lab is the first step toward this goal. It is a two-part lab to increase your environmental awareness and foster your desire to learn more about the environment, environmental problems, and solutions. The first part of the lab is to experience and observe phenomena of the natural world. The second part is to conduct basic library and Internet research on the environment. Observing environmental phenomena is an important component in environmental science; it is the first step in the **Scientific Method**.

MATERIALS

- ☐ Compass
- ☐ Measuring tape
- ☐ pH meter
- ☐ String, 25 meters
- ☐ String level
- ☐ Topographic map of campus
- ☐ Pen
- ☐ Notebook
- ☐ Internet access
- ☐ Library

TASKS

Part 1, the experiential exercise occurs outside (note there are fall and winter semester options depending on outside conditions) and Part 2, library/Internet research, occurs indoors. Be sure to credit the sources for your information.

WRITE-UP

This lab is an informal write-up in which you answer the following questions.

I—EXPERIENTIAL EXERCISE[2]

In designated teams, answer the following questions.

A. Go to a designated spot outside of the building:

   1. What are the general, ambient (background) weather conditions?

   2. How windy is it? There are lots of ways to estimate wind speed. You could drop a piece of fuzz and see if it drops or drifts, or look at trees, smoke, or a flag. Smoke rises straight up, and a flag droops straight down if there is no wind. If the wind is between 1 and 4 miles per hour (mph) the flag only occasionally flips open and the outer end hangs lower; smoke drifts to the side. Between 4 and 8 mph (a light breeze), the flag stirs more noticeably. Eight to 13 mph is a gentle breeze that moves branches. See if you can determine a reasonable estimate by looking at clues. Be sure to give both English and metric units.

   3. From which direction is the wind blowing?

   4. Are there clouds? What methods could you use to estimate their height from the ground? How high would you guess them to be?

   5. Select a sloping area. Using the string, string level, and measuring tape, determine the slope (percent and angle) over a 25-meter distance. Show your work.

   To calculate the average percent or angle slope, measure the distance (run) and the elevation change (rise). The rise divided by the run multiplied by 100 gives the **percent** slope. To determine the **angle** of the slope, divide the rise by the run; this gives you the tangent of the slope angle. Next, use a calculator or trigonometry table to take the arctangent or look up the corresponding angle. Figure 1.1 illustrates how to estimate slope.

[2]This exercise was adapted from L. Charles, J. Dodge, L. Milliman, and V. Stockley. 1981. Where Are You At?—A Bioregional Quiz. Coevolution Quarterly 32:1.

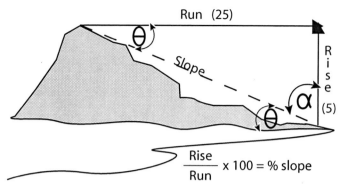

$$\frac{Rise}{Run} \times 100 = \%\ slope$$

For example: $5/25 \times 100 = 20\%$ slope

% slope	∠ slope
20 %	11.3°

$$\alpha + \theta = 90°$$

$$Tan\ \theta = \frac{opposite}{adjacent} = \frac{5}{25}$$

$$Tan\ \theta = .2$$

$$\theta = 11.3°$$

Figure 1.1—Illustration of Sample Slope Calculation.
Divide the rise by the run and multiply the result by 100 to get the percent slope. The slope angle is determined by dividing the rise by the run to get the tangent of the slope angle. Next, use a calculator or trigonometry table to take the arctangent or look up the corresponding angle in a table of tangents.

6. In what direction (compass direction) is the ground sloping?

7. As you look at the degree of slope, what are a few land uses that might be clearly *inappropriate*?

B. Go to a designated spot on the opposite side of the building

8. What is the wind speed? If it is different than question 2, explain why.

9. From which direction is the wind blowing? If it is different than question 3, explain why.

10. Select a slope. Using the string, string level, and measuring tape, determine the slope (percent and angle) over a 25-meter distance.

11. In what direction (compass direction) is the ground sloping?

12. In the nearest parking lot, look for the largest and oldest vehicles.

   a. What is the oldest vehicle (e.g., make, model, and year)?

   b. What is the largest vehicle (e.g., make, model, and year)?

   c. Which of these vehicles most likely consumes the most fuel? Reference?

   d. Which of these vehicles pollutes the most (e.g., tailpipe emissions, leaks from engine and transmission)? Reference?

13. Based on your campus tour, how many human-made devices did you see that increase or enhance wildlife (e.g., habitat, water, birdhouses, perches, wildlife cover)?

14. What is the benefit of doing a problem set such as this for environmental awareness?

Winter/Spring Semester Option (for areas that have snow)

Go outside your building to a location designated by the lab instructor.

15. Determine the mean depth of the snow (take at least three measurements to obtain the mean).

16. Using the snow sample, calculate the amount of water contained in the snow covering one acre. Determine the size of your campus and then calculate the amount of water in the form of snow on your campus. (Hint: To help you determine if you are on the right track, a rule of thumb is that heavy, wet snow can contain an inch of water in four or five inches of snow.) Give the amount in gallons and in liters. (See Question 17 in the next section for water quantities.)

17. Obtain a sample of snow. Bring it to the lab to melt during the rest of your time outside.

   a. How much water does it contain?

   b. What is the pH of the snow?

   c. Would it be defined as acidic (i.e., as in acid precipitation)? Reference?

18. What other contaminants/pollutants are likely to be in the snow on the campus? What are the sources of these pollutants?

19. When the snow melts on the campus, where does it flow? (Use your answers from the questions above to help you answer this.)

20. Provide the name, location, and distance to the nearest surface water body to the campus.

21. Based on your previous answers, what are the likely environmental impacts of the snowmelt?

22. Based on your experience outside, what species of trees or shrubs is the most prevalent? Are they native to this area?

23. How many different species of animals (e.g., birds, reptiles, amphibians, insects, and mammals) did you see or hear and what were they?

Fall Semester Option

Go outside your building to a location designated by the lab instructor.

15. Find a grove of trees or an urban forest edge or get as close as you can to one. Stay there for at least five minutes.

    a. Describe where you are and how to get there from the classroom (be specific).

    b. Glance around. How many different species of animals (e.g., birds, mammals, reptiles, amphibians, and insects) do you see or hear?

    c. Make a circle using your two thumbs and index fingers (about 10 centimeters in diameter) and place it on the ground over the grass, then count how many different plant and insect species you see in the circle. How many plants did you see? How many insect species? Was this more or less than you expected?

16. Find a new location that has no trees and repeat the steps above.

    a. Location:

    b. Species observation:

    c. Circle assessment: Did you find more or less than you expected?

17. How many gallons of water fall on a flat one-acre campus in the form of a 0.6-inch rainstorm? Convert this to liters. (Hint: 1 cubic inch = 0.017316 quarts, 144 square inch = 1 square foot, 1 acre = 43,560 square foot.)

18. A certain amount of the precipitation will infiltrate into the ground, some may evaporate, and the rest will run-off as stormwater run-off. (See Figure 1.2 for an illustration of precipitation infiltration and run-off.)

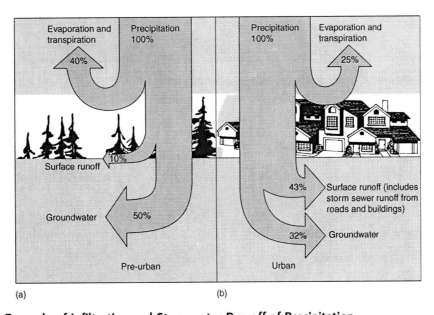

(a)   (b)

Figure 1.2—Example of Infiltration and Stormwater Run-off of Precipitation.
Urbanization can have a significant effect on the infiltration and stormwater run-off as urbanization increased impervious surfaces. Source: Raven, P.H. and L.R. Berg. 2004. Environment. 4th ed. John Wiley & Sons, NY (p. 285).

    a. What factors influence the degree of ground infiltration?

    b. Estimate the percentage of your campus that has impervious surfaces (parking lots, sidewalks, roofs, etc.).

19. On the campus, in which direction will the stormwater run-off flow? (Use your previous answers to help you.)

20. Figure 1.3 shows common pollutants in urban stormwater run-off. What pollutants (potential contaminants) are likely to be in the run-off from your campus? What are the sources of these pollutants?

Figure 1.3—Common Pollutants in Urban Stormwater Run-off.
Source: Raven, P.H. and L.R. Berg. 2004. Environment. 4th ed. John Wiley & Sons, NY (p. 499).

21. Using a topographic map, what is the name, location, and distance to the nearest surface water body to the campus?

22. Based on your previous answers, what are the likely environmental impacts of the stormwater run-off?

23. Locate a lawn near the classroom. Does it have nice green grass? Lawns are referred to as "nature on a leash."

    a. What does the term "nature on a leash" mean?

    b. What is required to keep "nature on a leash" and what are the environmental impacts?

    c. If maintenance were completely halted, what would the lawn look like in 5 years? 10 years? 25 years?

    d. Make a sketch to show what the abandoned lawn would look like in 50 years.

LAB PART II—LIBRARY/INTERNET ENVIRONMENTAL RESEARCH

The following questions are designed to further foster your basic environmental perception and awareness. For each answer, cite the reference.

1. What time did the sun rise today (date and time)?

2. When are the longest and shortest days of the year, _and_ what are the names for the events associated with these days?

3. When are day and night the same length, _and_ what is this phenomenon called?

4. How long is the growing season here? (Give approximate start and end dates.)

5. What was the total rainfall in the area last year?

6. Name three endangered mammals listed in this state.

7. What is the state's longest river, _and_ what is its length in kilometers?

8. When is the fishing season open for Atlantic salmon (_Salmo salar_)? (Or other species designated by the instructor:_____.)

9. Name four species of trees in this area (including the scientific name) _and_ identify which ones are native.

10. Is there a state official under the Governor in charge of both natural resources _and_ environmental protection? If so, what is that person's name and title? If not, what are the names and titles of the highest officials for natural resources and for environmental protection?

11. What is the source of your drinking water?

12. Where does your garbage go (i.e., name the facility and its location)?

13. What is the primary fuel source for electricity generated in your area, _and_ what are the major pollutants associated with this source of energy?

14. Determine how far away you are from the nearest:

   - National Forest
   - National Park
   - National Monument
   - National Wildlife Refuge
   - Local environmental recreation area

15. In the continental U.S., what and where are the tallest mountains east and west of the Mississippi River?

Science and the Media

OBJECTIVES

- Be able to list major challenges of communicating science in the media
- Know how to use basic library research methods for finding environmental science reference materials
- Be able to describe and critique several major environmental science controversies reported in popular print media
- Be able to work as a team in evaluating environmental information

KEY CONCEPTS & TERMS

✓ Research
✓ Peer review

INTRODUCTION

A critical component of environmental science is to identify, locate, and retrieve scientific information—the process of **research**. Research is essential to help you find a topic or narrow down the topic. You will need to research other studies that have been conducted because their results can assist you in formulating a hypothesis and designing an experiment.

Based on your research, you must evaluate the validity of the information source that you plan to use in an academic assignment. Print sources (e.g., books and articles found in the library) traditionally go through an editorial process that involves editors and fact-checkers verifying the information. The most reliable information is from peer-reviewed journals. **Peer review** is a process used for checking and verifying the work performed by one's equals—peers—to ensure it meets specific academic criteria. However, with the Internet, peer review is still generally not the case despite the increasing numbers of reputable on-line peer review journals (for example, *Environmental Health Perspectives* at http://ehp.niehs.nih.gov/). Anyone with a computer and access to the Internet can publish a web site. Thus, there is no guarantee that the information is accurate or true. Therefore, exercise extreme caution and employ skepticism when using information from Web sites to do research. However, you should also be cautious with print media. Using a book from 1975 as your primary source for a research paper on the health of the Atlantic salmon fishery is bound to lead to erroneous conclusions and a poor grade. (Salmon were listed as an endangered species in 2000.) Or sole reliance on a book published by the Coal Coalition or Greenpeace for your paper on the future of fossil fuels is likely to skew your conclusions. In both these examples, these books may serve as excellent supporting references, but only when combined with more current or neutral-position sources.

For many people, the popular media is the only source of information on scientific breakthroughs and environmental issues. One role of the media is to transfer knowledge from the scientific community to the public. The media releases what it considers to be of general interest to its readership from what it considers a "credible source." The media also focuses on frontier science (that which is not widely tested or accepted) because its breakthroughs make good news stories. Although newspapers and TV may be valuable sources of information, news stories generate confusion by using bureaucratic terminology unfamiliar to the public, by introducing bias, and often flagrantly appealing to the emotions rather than the intelligence of the public. Moreover, newspaper reporters and editors generally are not trained scientists, but journalists, which can affect their ability to distill and report science.

Environmental issues tend to be emotionally charged because they can influence human or animal health and welfare and can involve large sums of money. Controversy makes interesting reading—it may be profitable to inflate or focus on the extremes as a way to charge an issue and make controversies seem greater than they are. Journalists may also give both sides of an issue equal weight even if they do not have equal evidence. Consumers of public information need to be critical thinkers. This is especially true when the basic science is so young (try looking for Environmental Science as a term in 19th century literature).

The purpose of this lab is to evaluate the role the media plays in presenting environmental information and in promoting environmental awareness. Specifically, we will focus on the positive and negative contributions of newspapers.

MATERIALS

No special materials needed for this lab.

TASKS

- You will be divided into small teams to research print articles on recent environmental controversies and scientific breakthroughs.

- Each team will investigate one of the following topics (only one topic per team): West Nile Virus, genetically modified food, radioactive waste, national energy policy, forest fires, sewage sludge management, wilderness preservation, urban sprawl, MTBE, or suggest your own. Each team must find two newspaper articles (at least one has to be a national newspaper—Wall Street Journal, New York Times, Los Angeles Times, Washington Post, USA Today, etc.), two news magazine articles (e.g., Newsweek, Time, U.S. News & World Report), and one Internet site. Find print versions of the articles; if you get them from Web sites, photographs and tables might be omitted.

- As a team, evaluate the positive and negative contributions of the media coverage. Focus on the media's ability to impart the information necessary for the public to make properly informed decisions on environmental issues. Make sure to assess the ultimate source of the information, recognizing that the information may be biased because of vested or emotional interests. Your ability to evaluate information is a key to completely understanding environmental concerns.

■ At the beginning of next week's lab, a spokesperson from each team will give a short summary (5 min.) of the team's impressions and conclusions.

■ Each student will submit a typed report that answers the following questions (be sure to turn in copies of the articles).

QUESTIONS

1. What is your topic?

For each article:

2. What are the titles of your five articles? Be sure to provide the complete and proper citations. (See Part 1 for proper citation format or use one recommended by your instructor.)

3. What are the qualifications and expertise of each reporter?

4. Did you infer this or was it stated (as in J. Smith has a M.S. in molecular biology)?

5. What are the qualifications/expertise of the expert opinions cited or quoted (do this for each person or organization)?

6. Summarize and discuss the manner and degree to which vested (e.g., economics, jobs, health) or emotional interests are addressed in each article.

7. Discuss use of photographs, tables, figures, and other graphics. Were they accurate? What was the source? Did they support the thesis of the article?

8. Could you tell the writers' attitude towards science? How?

9. Did the articles in any way change your perception of the issue? Why or why not?

Overall

10. What do you think is needed for the public to have a better understanding of science?

Field Trip: Greening of Business

OBJECTIVES

- Be able to research and discuss the concerns connected with "greening" (environmental sustainability efforts) in a manufacturing plant
- Be able to describe the manner in which a "green business" is balancing economics and environmental concerns

INTRODUCTION

"Green" is an up-and-coming term you hear as corporations respond to environmental concerns. But becoming "green" for a manufacturer does not mean just making "green" products. Rather, it means addressing the "cradle-to-grave" aspects of the product by creating **environmentally sound product life cycles**. "Cradle" refers to the very start of the raw materials that eventually form the product, and "grave" refers to the ultimate environmental fate after the product has ended its useful life, having exhausted all recycling, re-use, and other options. Major factors to consider in life-cycle design include materials, energy, transportation, and toxicity.

Thus, ecological accounting principles can be applied to the world of commerce. "Green" businesses are beginning to grow. A Vermont company, Ben & Jerry's Ice Cream, made a name for itself in the 1980s and continues to thrive despite having to make some concessions to the larger corporate world. Tom's of Maine has become known as a proponent of natural health-care products and ethical corporate behavior. Ray Anderson, CEO of Interface, Inc., has gained international recognition as a result of his efforts to take the company toward complete sustainability. He argues that it is not just the right thing to do, it is also the smart thing to do; Interface's profits show the benefit of its greening efforts. Many areas have developed green business certification programs or green company associations. For example in California, the Bay Area Green Business Program was developed by local governments in collaboration with U.S. EPA, state government, and the business community to support greening and sustainability efforts; over 300 businesses and public agencies have been certified since 1996.

States with a "green" or socially minded business organization are becoming more common. In Maine for example, an environmental ethic is inferred through membership in Maine Businesses for Social Responsibility (MEBSR). The goal of MEBSR is to teach businesses the art and science of balancing economic success with environmental and social responsibility. Both large and small companies benefit form such efforts. The Center for Small Business and the Environment (June 2004, http://www.geocities.com/

aboutcsbe/) cites a recent study by the Aspen Institute's Nonprofit Sector Research Fund that found the National Federation of Independent Business (NFIB) is one of two nonprofit groups most cited by congressional staffers and members as having greatest influence on environmental issues (the other group is the Sierra Club).

As you address the following tasks, keep these questions in mind: What does it take to have a socially responsible company? How do we recognize these companies?

TASKS

- Research a company that would be suitable as an example of a "green" socially responsible corporation.

- Either visit this corporation individually or work with the instructor to arrange a class tour. Or set up a virtual visit by finding a company on the Internet and establish direct communications with the management of the company.

- Write an essay based on your research and virtual or actual field trip.

QUESTIONS TO ADDRESS IN RESPONSE TO THIS FIELD TRIP

1. Why was this company selected? What are its characteristics?

2. What environmental factors are involved in the day-to-day operation of the plant/offices?

3. What emphasis is placed on the health and safety of the workers?

4. How does this company deal with other companies in terms of corporate philosophy, marketing and product development, and environmental accountability?

5. How does a business such as this incorporate environmental accountability, yet still make a decent profit?

6. What should other companies be reasonably expected to do?

7. Where do the materials and services come from? What attention is given to this?

8. What happens to the product and packaging? (How is recycling and solid waste addressed?)

9. What does this company recommend for good environmental practices?

10. What supports are available in your state for "green" manufacturing and environmental sustainability for business and government (be specific)?

References

Bay Area Green Business Program. 2004. Home Page [Online]. Available at http://www.abag.ca.gov/bayarea/enviro/gbus/gb.html (verified 3 June 2004).

Business for Social Responsibility. 2004. Home Page [Online]. Available at http://www.bsr.org (verified 3 June 2004).

Center for Small Business and the Environment. 2004. Home of CSBE [Online]. Available at http://www.geocities.com/aboutcsbe/ (verified 3 June 2004).

Environmental Site Inspection

OBJECTIVES

- Be able to visually survey and inspect the buildings, structures, and grounds to determine if there are indicators of health and environmental risk, harm, or damage
- Be able to complete an environmental inspection form similar to those used in Phase I Site assessments
- Be able to describe procedures and observations from your site assessment

KEY TERMS & CONCEPTS

- ✓ Environmental Site Assessment
- ✓ Phase I Site Assessment
- ✓ Phase II Site Assessment
- ✓ Phase III Site Assessment
- ✓ Superfund

INTRODUCTION

Many concepts you learn in environmental science involve the prediction or evaluation of environmental impacts. In this lab, you will go to the field and try to identify actual health and environmental impacts by conducting a site inspection using a standard site inspection form.

Site inspection forms are used by environmental professionals to assess the potential for contamination and/or negative environmental conditions primarily for clients interested in purchasing property. That is, because of potential liability concerns, prospective purchasers of property want to know if there are potential or actual environmental problems.[3] This is done through a **Phase I Site Assessment**, which is also known as an **environmental site assessment**. The Phase I includes a review of historical records, federal and state files, and a visual inspection to look for actual or potential environmental problems (ASTM, 2003). If environmental problems are observed or are likely, the next step is a **Phase II Site Assessment**, which seeks to verify the observations and delineate the problems through sampling and analysis (ASTM, 2002). If the Phase II confirms the presence of significant problems, a **Phase III Site Assessment** is employed to remediate the problems.

[3]**Superfund**, officially known as The Comprehensive Environmental Response, Compensation, and Liability Act, establishes strict liability for property owners where hazardous substances present a threat to human health or the environment, regardless of when the hazardous substances were placed there.

TASKS

The tasks for this lab are: conduct background information on your site, inspect the site and complete an inspection form, and summarize the procedures and findings in a cover memorandum.

Task 1:

Typically, you will conduct some preliminary research on the property before the actual inspection. This research will include the examination of aerial photographs, Sanborn Fire Insurance Maps (see Lab 20), USGS topographic maps, National Wetlands Inventory maps, property deeds of adjacent landowners, and interviews with former owners and/or workers. In addition, you need to review government records pertaining to the site such as U.S. Environmental Protection Agency's Enviromapper (http://www.epa.gov/enviro/wme/) to identify Superfund sites, permitted water discharges, permitted air discharges, hazardous waste generators or management facilities, and facilities reporting toxic releases. This is done for the property of interest and adjacent and nearby properties.

For this lab, use the Enviromapper to identify onsite and adjacent areas of potential concern.

■ Input the site's zip code.

■ In the "Window to my Environment," check all the boxes of interest. To learn more about the items, click them for additional information.

■ Click "redraw map".

■ In the "Your Environment" box to the right of the map, click on the number next to "No. of facilities reporting to EPA" to obtain a list of facilities and addresses.

   1. What facilities did you find within one (1) mile[4] of the site?

   2. For each facility, list the regulated activity (air emissions, hazardous waste, water discharges, etc.)

   3. Is the facility of potential concern to your property? Why?

Task 2:

Go to the site designated by your instructor and complete the following form. If the site is a university building, use the approximate environs as the parcel, not the entire campus. Do not leave blanks or skip any items. If the item is not applicable, mark it "NA"; if an item was not inspected for, mark it "NI" and be sure the reader knows why it was omitted.

References

American Society for Testing and Materials (ASTM). 2003. E1527-00 Standard Practice for Environmental Site Assessments: Phase I Environmental Site Assessment Process. West Conshohocken, PA.

American Society for Testing and Materials (ASTM). 2002. E1903-97 Standard Guide for Environmental Site Assessments: Phase II Environmental Site Assessment Process. West Conshohocken, PA.

[4]This is an exception to the normal rule of using the metric system rather than English because the general public relies upon the EPA information at this site and English units are used in dealing with the general public.

ENVIRONMENTAL INSPECTION FORM

Date of inspection_____ Inspected by_____

Street and mailing address for the inspected property:

_____

_____

General description (e.g., 1/4 acre residential lot, manufacturer, commercial business, etc.)

_____

Legal documentation: Give Book and Page for the property's recording in the Land Records or tell where the property deed can be found. _____

_____

Weather conditions (precipitation, temperature, sun, wind, etc.)

_____

_____

Client: (in this case, your instructor) _____

Property Description and Address

Current Use of Property: _____ Residential: _____ Commercial: _____ Industrial: _____

Agriculture_____ Forestry _____ Other _____

Amount of raw (undeveloped) land _____

Past Use of Property: _____ Residential: _____ Commercial: _____ Industrial: _____

Agriculture: _____ Forestry: _____ Other: _____

General Field Observations

WARNING: The purpose of this lab is educational. Therefore, your health and safety are of paramount importance. Do not under any circumstances open, touch, move, or otherwise manipulate any unknown containers, articles, tanks, and so forth. If you observe such items, mark it as such. If you have a concern, bring it to the attention of the instructor. *Avoid contact!*

After reading the above statement, sign below that you understand the importance of health and safety for this lab.

Signature: _____ Date: _____

OUTSIDE INSPECTION (comment on all YES items)

Underground Storage Tanks (USTs)	YES	NO	UNK
Known or observable underground storage tanks? How many? ____			
Any evidence of underground storage tanks? ■ Fuel pumps ■ Vent tubes ■ Hoses ■ Manhole covers			
Any evidence of soil or groundwater contamination?			
Comments			

Aboveground Storage Tanks (ASTs)	YES	NO	UNK
Known or observable aboveground storage tanks? How many? ____			
Any evidence of aboveground storage tanks? ■ Signage ■ Berms ■ Pipes ■ Heating oil			
Any evidence of soil or groundwater contamination?			
Comments			

Drums/Containers	YES	NO	UNK
Known or observable drums/containers? How many? ____			
Any evidence of soil or groundwater contamination?			
Comments			

Ecological Conditions	YES	NO	UNK
Wetlands (onsite or adjacent) How large? ___			

Floodplains Location:			
Unique or critical habitat			
"Threatened" or "Endangered" Species			
Brooks/streams/rivers			
Ponds/lakes			
Earth disturbance			
Vegetation stains, matting, discoloration			
Comments			

Miscellaneous Conditions	YES	NO	UNK
Electrical power lines			
Discarded tires			
Onsite sewage disposal			
Abandoned wells			
Mounds or other potential signs of land disposal			
Building foundations			
Electrical equipment Labels indicating PCBs or no PCBs _____			
Comments			

INSIDE INSPECTION (For purposes of this report, do not conduct inside inspection unless authorized by lab instructor)

Lead	YES	NO	UNK
House built before 1978?			
Any evidence of lead-based paint?			
If yes, is paint chipping, cracking, or peeling?			
Evidence of lead piping for water?			
Comments			

Asbestos	YES	NO	UNK
Evidence of asbestos (insulation, pipe insulation, tiles, and siding)?			
Comments			

Hazardous Materials	YES	NO	UNK
Pesticides (herbicides, insecticides, etc.)			
Paint			
Fuels			
Unknown/Unidentifiable			
Urea formaldehyde foam insulation			
Comments			

Task **3:**

You need to communicate your findings to your client, in this case, the lab instructor. This will require you to prepare a cover memorandum with a copy of your inspection checklist as an attachment. Use the following format for your cover memorandum. (Be sure to attach a copy of the inspection checklist.) The tone of your memorandum should be neutral and fact-based.

MEMORANDUM

TO:
FROM:
DATE:
SUBJECT:

First Paragraph—Discuss what you did, when you did it, why you did it, and where you did it.

Second Paragraph—What did you find and what did you not find.

Third Paragraph—Recommendations (e.g., "No actual or potential problems or further investigation is needed because. . . .")

Signature

Attachment—Attach a copy of your inspection checklist.

Life Cycle Assessment

OBJECTIVES

- Be able to describe the concept of *upstream* environmental impacts of consumer products
- Be able to describe the components of Life Cycle Assessment (LCA)
- Be able to present a cradle-to-grave assessment for a particular product and describe how the analysis is used for pollution prevention/reduction

KEY TERMS

- ✓ Life Cycle Assessment
- ✓ Pollution prevention
- ✓ Source reduction
- ✓ Upstream impacts

MATERIALS

- ☐ Internet access
- ☐ Poster board and art supplies

INTRODUCTION

All products have some impact on the environment. However, some products require more resources, use more toxic materials, emit more pollution, or cause greater impacts when disposed. To promote sustainability, we must design, produce, and use products that require less resources, energy, and toxic materials. We should produce products that have minimal impact when used, and that can be easily reused or recycled after their useful life.

Life cycle assessment (LCA) is a process to identify and evaluate the potential environmental effects of a product over its lifetime (EPA, 1993). That is, a "cradle-to-grave" analysis. Because you generally focus on a product, which is the end result of a manufacturing/processing effort, you need to work backwards—**upstream**—to examine the various impacts resulting from every activity required to produce, distribute, use, and eliminate the product. LCA is a powerful tool that can be used to assist public agencies to formulate environmental policies, help manufacturers reduce energy consumption and environmental impacts of their products, and help consumers make more informed choices.

To apply LCA, you must sketch the product's life cycle, identify resource inputs, identify outputs, assess the associated potential environmental impacts, and identify *pollution prevention* measures. (See side bar).

Conducting a complete and thorough LCA is complex, time-consuming, and expensive, as it requires detailed data and knowledge. However, a simplified, conceptual LCA can be produced relatively easily and provides important, basic information.

There are five major steps in applying LCA.

STEP 1: Create a flow diagram depicting the manufacture, processing, use, and disposal of the product.

STEP 2: Identify the major inputs (resources and energy) and types of raw materials and energy fuels.

STEP 3: Identify the major outputs (e.g., the useable product, hazardous and non-hazardous waste, and pollution), including the specific type of pollutant and waste.

> ### Pollution Prevention*
>
> Pollution prevention is based on *source reduction*, which means any practice that (a) reduces the amount of a pollutant entering any waste stream or otherwise released into the environment prior to recycling, treatment, or disposal; and (b) reduces the impact on public health and the environment from the release of such pollutants. Source reduction practices include equipment or technology modifications, process or procedure modifications, reformulation or redesign or products, substitution of raw materials, and improvements in housekeeping, maintenance, training, or inventory control.
>
> *Pollution Prevention Act of 1990, United States Code Title 42, Chapter 133, Sec. 13102.5(A)

See Figure 5.1 for an example of a conceptual LCA depicting the life cycle, inputs, and outputs.

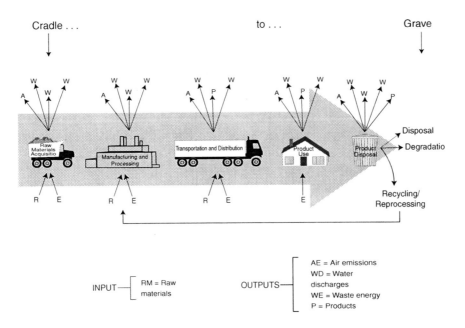

Figure 5.1—Conceptual Life Cycle Assessment with Inputs and Outputs.
Assessing the life cycle, from cradle-to-grave, of a product is becoming more common in manufacturing and other corporate and government processes.

STEP 4: Evaluate the potential environmental impacts of each stage. Based on the inputs and outputs, the impacts on human health and the environment can be determined.

For example, if the primary fuel source is coal, the environmental effects may include erosion/sedimentation and acid mine drainage from mining; production of acid deposition and greenhouse gases through combustion; and landfilling of resultant combustion ash. The manufacture and processing of the product may release volatile organic compounds, which are a principal component of ground-level ozone. The use of a product (e.g., automobile) may generate air emissions and result in spills to water and soil. The disposal of the product, when it is no longer wanted, may result in air emissions if it is incinerated, or releases to the soil and groundwater if it is landfilled (e.g., mercury battery).

STEP 5: Identify pollution prevention options for each energy source and environmental impact. For example, is there a cleaner energy alternative that can be used for the product? Can the product be manufactured with safer and less polluting substances? Can the product be designed with less packaging? Can the product be designed so that its emissions during use are reduced? Can the product be designed to increase its reuse or recycling potential so that landfilling or incineration can be avoided?

TASK

■ Select a consumer product with the help of your lab instructor.

■ Create a poster that depicts and documents the principles of LCA for your product. Your poster should be neat and professional in appearance and reflect effective communication principles.

1. Provide a detailed description of the product (i.e., dimensions, contents, use, and so forth). Include a photograph, drawing, or actual product. What is the quantity or mass of the product?

2. Using the Internet and library for research data, draw the product's conceptual life cycle on the poster as depicted in Figure 5.1 using a block flow diagram.[5] The life cycle should include the raw materials acquired, manufacturing/processing, transportation/distribution, disposal, and the environmental fate of the product and byproducts so you've captured the "cradle-to-grave" aspects.

   a. For each product stage, identify and list the input and outputs: For these steps, you will need to conduct some basic library or Internet research.

      i. List the likely energy source and raw materials used to produce the product (including its packaging).

      ii. Where feasible, provide numerical data (e.g., x% fiber, x% mineral sprits, etc.)

      iii. List the likely emissions, types, and amounts of pollutants and wastes.

---

[5]A block-flow diagram is a schematic illustration of a major process. The block (or rectangles) represents a unit operation or step. These blocks are connected by lines or arrows that represent the process flow between the unit operations/steps. Many figures in this manual are block-flow diagrams.

b. For <u>each</u> product stage, identify and list the environmental impacts.

    i. For each pollutant, list the basic environmental/public health effect.

    ii. For each listed effect, properly cite the information source.

c. For <u>each</u> product stage, identify possible pollution prevention measures, which could reduce the overall environmental impact of the product, including production, transportation, use, and disposal.

d. Be sure to cite references properly and clearly on the poster.

Reference

U.S. Environmental Protection Agency. 1993. Life Cycle Assessment: Inventory Guidelines and Principles (No. EPA/600/R-92/245). GPO, Washington, DC.

Urban Ecosystems

OBJECTIVES

- Be able to describe an urban ecosystem and identify its major components
- Be able to describe an urban inventory approach used in planning
- Be able to identify energy pathways in a particular streetscape setting within an urban ecosystem

INTRODUCTION

The immediate environment for most people is an urban area. The urban environment is an ecosystem, and like any other, it contains units and pathways for the exchange of energy and information. The urban environment is studied by planners and policy makers, who are increasingly incorporating ecosystem management concepts in their professional work. Traditional biological and other environmentally related sciences also must consider the effects of urban environments; the effects of urban settlements and the industrial revolution can be felt everywhere, including remote Antarctica.

Your textbook provides a basic discussion and reference for various forms of pollution, infrastructure, social benefits, and environmental impacts associated with cities. In this field activity you have an opportunity to combine this information in an inventory and analysis of a block in a downtown area. This is similar to what might be done by planners, landscape architects, civil engineers, and other professionals involved in urban revitalization or related planning or economic development efforts. The gathering of urban environmental data for planning and design is complex and more formal than as is presented in this lab, but this lab will provide an introduction to the kinds of things considered in urban design and management.

In the interest of time, we will be conducting a condensed version of three activities. First, you will sketch the profile for an urban unit. Second, you will inventory the different types of energy pathways (infrastructure) entering and leaving the unit—driveways, sidewalks, doors, telephones, mail slots, power lines, smokestacks, exhaust vents, any connecting thing into air, land, water, or other medium. Third, you will estimate the units or range of units used to measure use of these pathways (e.g., number of cars using driveway, smoke going out of chimney, and so forth).

To put yourself in the right frame of mind, imagine that your company has been hired by the city to apply for an urban renewal grant, and you want to get a handle on conditions by conducting a preliminary non-intrusive walkover. In such a process, one might use transacts on a grid or by assessing socioeconomic or economic spatial sectors. (See Figure 6.1 for a model of an urban spatial structure functioning as an urban ecosystem.)

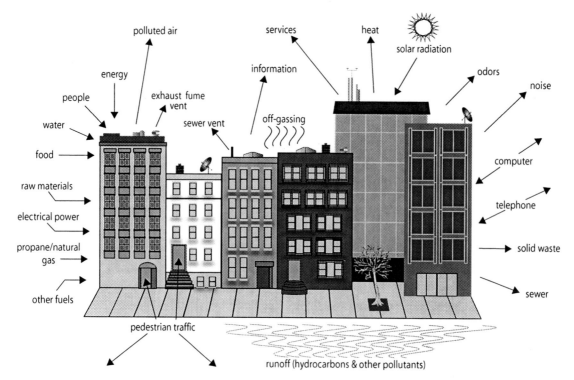

Figure 6.1—Urban Ecosystem Diagram.
An urban system exchange of energy and materials. An urban ecosystem contains at least an order of magnitude more energy per unit area than other ecosystems.

MATERIALS

☐ Hand-held counter (optional)

TASK

In teams, using a grid system, you will be assigned a "streetscape." Review Figure 6.1, the Urban Ecosystem. This shows a schematic system input-output. At your study site, sketch a profile of the streetscape scene (called a "street elevation") in your notebooks or on graph paper. A cartoon-like outline is adequate; the important things are the basic information and the concepts. Include buildings, fences, roads and driveways ("curb cuts"), signs, dumpsters, and anything else in the built environment, as well as grass, trees, and other natural features. The cross-section should include the ground and what you think is below the ground (water, sewer, storm water, electrical, telephone, gas, etc). Include the roadways on either side of the block—this is where your study site may overlap with that of another team. If you are not sure whether to include something or not, put it in; it is much easier to take it out later than the other way around.

1. Describe your block setting so that someone from away can find the town, the street location, and the block face you studied.

2. Inventory the structures and features you see.

    a. What kinds of buildings are there?

    b. What types of businesses are there? How many employees or families work/live there?

    c. Present the results in a table.

    d. Describe your interpretation of the evolution or development of the setting (streetscape).

    e. Can you tell the age of the buildings?

    f. What changes have the buildings undergone?

3. What inputs and outputs do you see? Refer to the attached figure, *major inputs and outputs of an urban area*, but use your own judgment on how and what to label. Draw and label arrows on your diagram. To help you get started, you can use the following table for ideas about identifying and labeling atmospheric inputs and outputs

Potential types and sources of urban air pollutants

POLLUTANT	EXAMPLE, SOURCE (INPUT)
Carbon monoxide	Vehicles, industry using oil or gas, building heating systems, parking lot
Sulfur oxides	Industry using coal and oil, heating using coal and oil
Nitrogen oxides	Vehicles, heating buildings (space heating) using oil and gas
Hydrocarbons	Vehicles, burning processes, fuel station
Particulates	Vehicle exhausts, industry, building heat, open burning, spore & pollen-bearing vegetation
Odors	Chemical and industrial processes, vehicles, food processing, solid waste, stack discharges, numerous others
Noise	Vehicles, people, industrial, and manufacturing processes
EMR/EMF	Electrical motors, transmitters, transformers, and other electrical and broadcast equipment

You should also look at water inputs and outputs. You can make a table for this too:

Potential types and sources of urban water pollutants

POLLUTANT	EXAMPLE	SOURCE OF INPUT TO SYSTEM
Storm water	Drainage ditch	Open transport away from site [etc.]

4. What other inputs are there? If these are hard to think of, refer to the figures, or list the outputs and work back from them. Describe each input and output. Estimate the units for each input and output. Can the input be compared with the potential or available sources?

5. Determine the total number of motorized vehicles on your street for a one-hour period. Do this by counting the vehicles driving in one direction (the nearest lane or lanes) for a half hour period. Multiply the number by four to covert the figure to an estimate of what you might expect if you counted for an hour and to incorporate cars traveling in the other direction. Be sure to record the exact time of the day you did your count. When do you think the peak travel hours are in the morning and evening? How many vehicles do you estimate are on the road during those times? Comment on the capacity of this urban area to accommodate this traffic.

6. Vehicles may enter your site (including parking in front) in generally one of two manners: by design ("destination" trips) or by "capture" (impulse, wrong turn, or advertising caught driver's eye, or something similar).

   a. How many vehicles entered?_____. How many vehicles exited?_____.

   b. Did any pedestrians enter the buildings or green space during the half hour you were observing traffic? If so, make a table:

   Sample table

PEDESTRIAN DESCRIPTION	TIME
Male shopper	3:15 PM
Female child, entered residence	3:20 PM

To do estimates for sewer and water, we could simply read meters (DO NOT DO THIS). Or we could use billing records at the municipal offices (again, do not do this yourself). If possible, you can estimate the size of sewer, water, and other utility pipes. If you cannot tell without getting intrusive, simply label as residential or commercial or other category based on apparent use.

7. Estimate noise from the site by subjective description (e.g., heard hum of motors, heard shouting). Use a table to present this information, but you do not need to record the time.

8. Describe any odors you might detect and record this along with information you might have on possible sources. Example: No cars in sight, but smelled exhaust from Main Street. Again, you can use a table to record this information if there is a variety, but you do not need to record time.

9. Are there different land uses in your site, such as houses on one end and stores on the other? If so, the boundary should be labeled on your sketch. Decide if other types of spatial information should be included, such as those based on socioeconomic status, as in working-class apartment building next to middle-class house. Of course, these are just preliminary impressions, but they serve to articulate a *sense of place*—part of the "understanding the problem" phase in environmental planning.

10. Now that you have spent a little time examining an urban ecosystem, what do you see in common with other forms of ecosystems? How has this changed the way you view an urban setting?

WRITE-UP

This is a professional report, not a lab experiment, but should be written in a formal manner. Any tables you create should be typed, along with a brief explanation of how those tables were derived. Your methods and results should be included, along with reflections on any benefits and limitations you experienced in carrying out the field activity.

Experimental Design: Law of Tolerance

OBJECTIVES

- Be able to describe an environmental application of the scientific method
- Be able to design and carry out an experiment on how abiotic environmental factors affect organisms
- Be able to describe the role of data collection and interpretation in environmental science

KEY CONCEPTS/TERMS

- ✓ Abiotic factors
- ✓ Biotic factors
- ✓ Experimental design
- ✓ Habitat
- ✓ Hypothesis testing
- ✓ Range of tolerance
- ✓ Scientific method

INTRODUCTION

Species vary in their resource needs and tolerances. For example, through evolution, some species thrive where it is hot (e.g., desert dwellers) while others thrive in the cold (e.g., Arctic species). The environment consists of **biotic** (living) and **abiotic** (nonliving or physical) parts. Within a given ecosystem, whether it is the Mojave Desert or the Arctic, each population of a species has a **range of tolerance**—the environmental conditions a species can tolerate. Individuals within a population may also have slightly different ranges of tolerance for any environmental condition. An organism's size, age, state of health, or genetic code can influence its tolerance range. For example, very old or young roadrunners (*Geococcyx californianus*) in the desert may be less able to tolerate extremely hot temperatures than mid-aged individuals. But there is also a limit when it is too hot for any roadrunner. (See Figure 7.1 for a sample display of a tolerance range.)

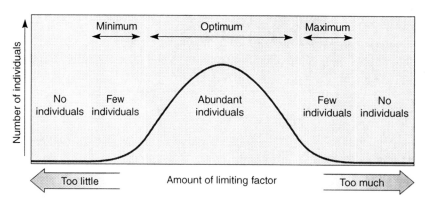

Figure 7.1—Limiting Factors and the Law of Tolerance.
The limiting factor for a species in an ecosystem is any environmental resource present in excess of an organism's tolerance or in insufficient quantities to meet the organism's basic needs. Source: Raven, P.H. and L.R. Berg. 2004. Environment 4th ed. John Wiley & Sons, NY (p. 96).

The individuals within a population of a given species respond to certain factors of their environment and seemingly ignore others. Many characteristics of a **habitat** (local environment in which an organism, population, or species lives) are variable from time to time or at different locations within the habitat. Temperature, quantity of light, and wind speed often vary in a forest habitat. When specific environmental factors vary continuously over a distance, a gradient exists. Light intensities can range from absolute darkness to extreme brightness. A shady spot may be a few degrees cooler than a position in direct sunlight only a few meters away. Even at a micro level, each organism has optimal conditions to thrive.

In this exercise, using the **scientific method** (see Figure 7.2), you will produce an appropriate **experimental design** (creating an experiment through which a hypothesis can be tested) to determine if fruit flies (*Drosophila melanogaster*) prefer certain light or heat conditions. You will design the experiment using the equipment and apparatus provided, carefully collect data, and analyze the data to determine which environmental variables are significant to fruit flies and how the flies are influenced by key abiotic gradients.

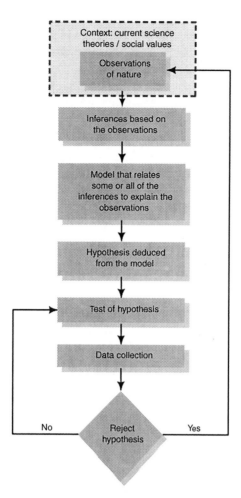

Figure 7.2—The Scientific Method.
Essentially, science is a disciplined way of asking questions. Feedback in this system supports the self-correcting progress of scientific information. Source: Bodkin, D.B. and E. A. Keller. 2003. Environmental Science: Earth as a Living Planet. 4th ed. John Wiley & Sons, NY (p. 22), modified from Pease, C.M. and J.J. Bull. 1992. Bioscience 42:293–298.

MATERIALS

☐ Aluminum foil	☐ Light
☐ Beakers	☐ Thermometer
☐ Cardboard	☐ Light meter
☐ Duct tape	☐ Paper towels
☐ Fly vials	☐ Petri dishes
☐ Fruit flies	☐ Flexible and/or rigid plastic tubing
☐ Graduated cylinders	☐ Ring stand
☐ Hot plate	☐ Sponge stoppers
☐ Ice	

TASKS

During this exercise, the class will be divided into groups. Half the groups will work with light, and the other half will work with temperature to: (1) determine whether fruit flies are affected by abiotic factors, (2) devise a way to make this determination, and (3) construct a curve depicting the range of tolerance. The instructor will discuss some background fly facts and considerations for experimental design.

Steps for this experiment:

1. Develop a hypothesis about how flies are affected by your assigned variable (light or temperature). You will want to choose a range for the variable that will capture an effect (e.g., too hot a temperature will cook all the flies).

   Note: Bring your written hypothesis to the lab instructor. If approved, design your experiment for your assigned gradient factor. Remember, you are designing an experiment to test your stated hypothesis.

2. Design an experiment to test the hypothesis. The instructor will provide a variety of materials that can be used to construct an experimental apparatus to run your experiment. Other potential variables will complicate your experiment, so try to make them as constants so that they do not interfere. For example, if you are measuring the effects of illumination, you do not want the flies to be so cold that they cannot move toward or away from the light. Be sure you have a control set up as part of the experiment. The control should replicate everything except the variable you are manipulating.

   Bring your written experimental design to the lab instructor. After it is approved, you will be provided fruit flies to conduct your experiment.

3. Assemble and adjust your apparatus so that you establish the specific gradient assigned to you.

4. Place your flies in the apparatus.

5. Record background conditions—temperature, light, and other environmental variables as determined by the class or instructor.

6. Allow the flies time to move to their preferred position along the gradient. How much time do you think is reasonable? Be sure you keep notes on this and other factors for your lab report.

7. Collect data concerning population density at positions along the gradient.

8. Record your data.

9. Interpret the data.

10. Repeat the experiment two more times.

LAB WRITE-UP

A formal lab write-up (i.e., a report with title, introduction, methods and materials, results—including table or figure, and discussion) is required for this lab. Follow the guidelines from Part 1 of this lab manual. Be sure to make a diagram to show how your experiment is set up. In writing your discussion section, consider the following (i.e., do not answer these questions, *per se*, but use the following as a guide in discussing your experiment):

1. What possible confounding (interfering or overlapping) variables might make one trial different from the other (e.g., gender, age, stress)?

2. Discuss what variables might make the control yield different experimental results from the trials. And discuss to what degree you can control these variables.

3. If you were doing it again, discuss what changes you would make in setting up your experiment.

CHAPTER

EIGHT

Experimental Design: Environmental Contamination

OBJECTIVES

- Be able to design and conduct an environmental contamination experiment
- Be able to describe the potential adverse effects of contaminated soil on plants and how to measure these effects
- Be able to describe the use of hypothesis testing and related statistics in contamination studies

KEY CONCEPTS & TERMS

- ✓ Anthropogenic
- ✓ Contamination
- ✓ Experimental design
- ✓ Hypothesis testing
- ✓ Null hypothesis
- ✓ Phytotoxicity

INTRODUCTION

Contamination of soil from **anthropogenic** (human-made) pollutants is a widespread and serious environmental problem. Soil can become contaminated through a variety of human activities including the using, unintentional spilling, and intentional discharge of hazardous materials and waste; the engineered treatment and disposal of waste; and the deposition of air pollutants.

   In this lab, you will be examining the **phytotoxicity** (poisonous to plants) by simulating, testing, and measuring the effects of contaminated soil. You will be provided seed, soil, and a contaminant. Based on this, you will need to formulate an hypothesis, including the null hypothesis (see *Glossary*). Then, you will develop an experimental design to test your hypothesis.

MATERIALS

- ☐ Hard red winter wheat seeds (*Triticum aestivum*)
- ☐ Plant containers
- ☐ Water
- ☐ Soil
- ☐ Common environmental contaminants. For example:
  - ○ Table salt (or road salt mixture used for de-icing)
  - ○ Motor oil
  - ○ Antibacterial soap containing Triclosan
  - ○ Automatic dishwashing detergent containing phosphorus
  - ○ Isopropyl alcohol
  - ○ Other

TASK

In this lab, you will be provided basic materials, direction, and a description of the expected outcome; however, you will formulate your hypothesis and design the experiment with minimal assistance from the instructor. The procedures are as follows:

1. For each group, select a contaminant (only one per team) to test.

2. Formulate a hypothesis. Be sure the hypothesis is appropriate and testable and refers to the contaminant (e.g., x% of contaminant will . . .). Restate the hypothesis as a <u>null hypothesis</u> for this experiment.

3. Based on your hypothesis, design an experiment to test your contaminant's effect on the growth of 10 red hard winter wheat seeds (note that the seeds take 3 to 4 days to germinate).

4. Select four (4) treatments (four different concentrations to test). Select your treatments so that they bracket the effect you think might result. For example, your hypothesis states that 5% of your contaminant will cause the effect. To bracket this, you would test concentrations above and below this percent. If you do not do this, you might find that your levels of toxicity are all too low or too high.

5. Establish a control group.

6. Replicate the experiment twice (for a total of three times). Thus, you should have <u>15 containers</u> per team, and it should look like Figure 8.1.

7. Note that your experiment will take approximately 7–10 days. Thus, you are responsible for maintaining your experiment (i.e., do not forget to water your plants). Should your experiment fail for any reason prior to its completion, you will have to start over. You may have to readjust your concentrations—much of science consists of learning from our so-called failures, so do not be discouraged.

8. Test your hypothesis. For each trial, you are comparing the mean plant growth for <u>each</u> treatment to the appropriate control (i.e., for Trial 1, 1Ct to 1a, 1Ct to 1b, 1Ct, to 1c, and 1Ct to 1d. For Trial 2, 2Ct to 2a, 2Ct to 2b, 2Ct, to 2c, and 2Ct to 2d).

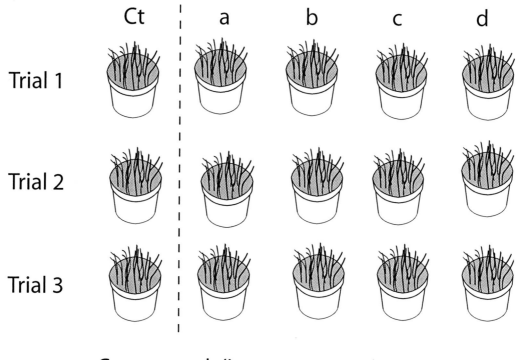

Ct = control (i.e., no contamination)

Figure 8.1—Experimental Design Set-up.
In this experimental design, three trials are conducted. Think of each trial as replicates. Thus, each trial must be the same and each trial must have a control.

WRITE-UP

This is a formal lab write-up:

- Title
- Introduction (including hypothesis)
- Materials & Methods
- Results (including tables or graphs)
- Discussion

In your discussion section, also relate your findings to real-world application. This will depend on your contaminant (e.g., road de-icing, oil leaks, household cleaners, sewage discharge into surface water, septic tank systems, groundwater, wildlife, fish, and so forth.)

OPTION 1—HYPOTHESIS TESTING

One of the most important elements in environmental science is hypothesis testing. That is, are your results significant because they enable you to accept or reject your hypothesis with a reasonable probability of being correct? This is done through statistics—using numerical data to make comparisons and conclusions—specifically tests for significance. This involves the application of a formal procedure for comparing observed data with a hypothesis. The results of the test of significance are expressed in terms of a probability that measures how well the data and hypothesis agree. Although we have not addressed statistics in this lab manual, you will only need to understand the basic concepts of hypothesis testing and Microsoft Excel. You will be conducting a statistical test to determine if your results are significant. That is, can you accept or reject your hypothesis?

T-TEST FOR INDEPENDENT SAMPLES

The *t*-test is the most commonly used test of significance to evaluate the differences in arithmetic means between two groups. For example, the *t*-test can be used to test for a difference in test scores between a group of patients who were given a drug and a control group who received a placebo (i.e., treatment v. control).

The *t*-test compares the mean of the two independent variables to determine if they are statistically different from each other. For this lab, example hypotheses are:

H_A: Seeds subjected to 10% of pesticide A will have a higher mortality rate.

H_0: There is no difference in mortality when seeds are subjected to 10% of pesticide A.

You need to generate a null hypothesis for this experiment.

Based on an experiment, the following dataset (see Table 8.1) was found. This can be analyzed with a *t*-test comparing the mean survival in control and treatment groups (or whatever measure you choose in your hypothesis). Remember, you must measure the height for all 10 seeds. If there are only 9 plants but you planted 10 seeds, the height for the non-germinating plant is 0 because you need to calculate the mean height. Ideally, most or all seeds germinate because the range of your variable is within what the plants can tolerate, enabling you to capture variation in response.[6]

TABLE 8.1 Data Table Example for Plant Testing*

Container 1	Plant	Treatment (Height of plants in cm)	Control (Height of plants in cm)
	1	5	7
	2	5	7
	3	8	10
Mean		6	8

*Sample data set for Trial 1, Treatment A.

By conducting the *t*-test, you are asking whether there is a statistically significant difference between the mean height of the treatment group as compared to the mean height of the control group. Assuming the only difference between the two is the treatment, you are assessing whether or not the treatment was responsible.

We will use Microsoft Excel to test the hypothesis. Using the data above, and the "two sample "t-test assuming equal variance," we get a "two-tail critical value of 2.776, a p-value of 0.01, and a "t-statistic of 2.12" (it is acceptable to drop the negative sign for this exercise). What does this tell you? For the purposes of this lab, we will use the p-value.

[6]If the treatment is likely to produce mortality, perhaps an alternative comparison might be to test whether or not the seeds germinate. If you use this method, your instructor might want you to plant at least ten seeds per treatment. Increasing the sample size helps us be more confident that measured differences in germination reflect the results of the treatment.

The p-value is a measure of probability that a difference between groups (treatment and control) during an experiment happened by chance. For example, a p-value of 0.01 ($p = 0.01$) means there is a 1 in 100 chance the result occurred by chance. The smaller the p-value, the stronger the evidence against H_o provided by the data. That is, the more likely it is that the difference between groups was caused by the treatment. Small p-values suggest that the null hypothesis (i.e., no effect) is unlikely to be true. Thus, the smaller the p-value, the more convincing is the rejection of the null hypothesis. Note that the p-value indicates the strength of evidence for rejecting the null hypothesis.

You will need to establish the significance level of the test, which for science generally is 0.05.[7] (This means that the data provides strong enough evidence that your result would occur less than 5% of the time—or 1 in 20—if the red winter wheat populations were really identical.) If the p-value is less than 0.05, which in this case it is ($p = 0.1$), you can state the following: "The results are significant at the 0.05 level that x% of x contaminant by weight had a significant effect on the (mortality, growth, height, etc.) of red winter wheat."

MICROSOFT EXCEL DIRECTIONS

You need to compare the data (i.e., the height) in the control and the data for each treatment in each trial. Thus, you will be running four t-tests for each of the three trials for a total of 12 t-tests. To run the *t*-test in Excel, after entering your data in columns (see Table 1), select "Tools," then select "Data Analysis," and then select "t-Test: Two-Samples Assuming Equal Variance."[8] For the Input, "Variable 1 Range" should be all the values in the control column. "Variable 2 Range" should be all the values in the treatment (the order does not matter). Then select OK. In the computed results, you want to check the p value. If the p-value is less than 0.05 (i.e., 5%), the result is statistically significant at the 95% confidence level. Reporting the p-values enables the reader to form his or her own conclusion about the significance of the results.

OPTION 2—SCIENTIFIC COMMUNICATION

The experimental design you prepared for the first part of this lab on environmental contamination will form the basis for a presentation to the class using Microsoft PowerPoint (or other format selected by your instructor). This is a group effort; each person must have some role in orally presenting the information. Your presentation should be no longer than 15 minutes in length and will be followed by a question and answer session. Your presentation should follow your formal lab write-up:

1. Introduction (your hypothesis and why did you choose this hypothesis)

2. Materials and Methods

3. Results (What happened? Use graphs, charts, and/or tables.)

4. Discussion (why did the results happen the way they did, do you accept your hypothesis, were there any flaws in your experiment?)

5. Conclusion (significance to environmental science)

---

[7] If you choose a significance level of 0.01, your result will occur less than 1% of the time—1 in 100—if the red winter wheat populations were really identical, which is obviously stronger.
[8] For some versions of Excel, you may have to install the data analysis component in order to run this *t*-test. Go to "Tools" and select "Add Ins." Check the boxes for "Data Analysis Toolpak" and then click OK.

Trophic Ecology of Humans:
The Best Guess Breakfast Interview

OBJECTIVES

- Be able to describe your own awareness of personal diet and food choices
- Be able to estimate the amount of energy required to provide food on the table

KEY CONCEPTS & TERMS

- ✓ Energy investment in food production
- ✓ Nutritional value
- ✓ Caloric content

INTRODUCTION

What we eat has a direct impact on the environment. As illustrated in Figure 9.1, the production, processing, storage, and distribution of food require energy, pesticides, fertilizers, and a variety of other inputs. Moreover, there are significant inputs and outputs in packaging, selling, and advertising food products. Energy to provide food can be estimated as a function of growing food, processing it, and transporting it to markets. If the market reflects the true cost of production, then the market cost is a good representation of energy, and even if it is not, it still represents the energy you took to earn the dollars. Food returns energy to us in the form of calories (a food **calorie** as used in product labeling and nutrition is actually the same as a **kilocalorie** as used in science). Nutritional value represents efficient calories that provide what your body needs to stay active and healthy.

Land that produces plant products yields greater kilocalories per square meter per year than land used to produce animal products. Animals are consumers, and as energy flows up to them from lower trophic level producers (plants), there is a loss of energy between the trophic levels through respiration, heat, and animal waste.

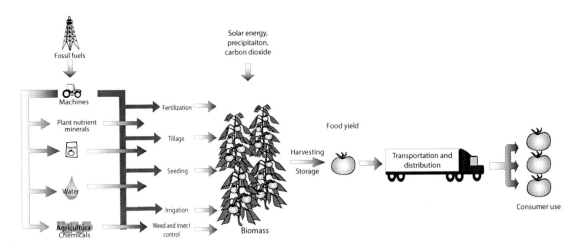

Figure 9.1—Inputs/Outputs of Food Production, Processing, Transportation, and Distribution.[1]

[1]Based on Figure 18.6 (Energy inputs in industrialized agriculture), In Raven, P.H. and L.R. Berg. 2004. Environment. 4th Ed. John Wiley & Sons, New York.

Plants can be divided into two groups based on their photosynthetic and respiratory pathways. C-4 plants are fast, efficient photosynthesizers compared to C-3 plants.[9] Cane sugar is a C-4 plant with a yield of 3500 kCal/m²/year;[10] corn cereal—another C-4—yields 1600 kCal/m²/year. Tables 9.1 and 9.2 show the yields of certain plant and animal food sources (Brewer and McCann, 1982).

TABLE 9.1	**C-3 Plant Categories and Their Yield in kCal/m²/year**		
Bread	650	Apples	1500
Wheat (& cereal)	810	Pears, peaches	900
Oranges, grapefruit	1000	Vegetable oil	300
Frozen OJ	410	Margarine	300
Peanut butter	920	Beet sugar	1990
Rice (& cereal)	1250	Coffee	12
Potatoes	1600	Tea	40
Other vegetables	200		

[9]Most plant species are "C-3" (or C3), so-called because they make a three-carbon compound as a stable product of carbon fixation. Plants that make a four-carbon product with PEP carboxylase are called "C-4" (or C4) and lose far less carbon through photosynthesis than do C-3 plants. Less than 1% of plant species are C-4.

[10]10,000 square meters equals one hectare or 2.47 acres. Caloric yields are just estimates for comparative purposes since many factors affect volumetric production as well as food quality. Data for Table 9.1 and 9.2 based on Akoroda, M.O. 1998 Comparative Output of Calories from Starchy Food Crops in Sub-Saharan Africa. Trop. Agric. 75:257-262; Brewer, R. and M.T. McCann. 1982. Laboratory and Field Manual of Ecology, Saunders College Publishing, Philadelphia; Smil, V. 2001. Enriching the Earth: Fritz Haber, Carl Bosch, and the Transformation of World Food. MIT Press, Cambridge, MA.; US Department of Agriculture, National Agricultural Statistics Service. 2004. Agricultural Statistics: An Annual Report. Available at http://www.usda.gov/nass/pubs/agstats.htm (verified 14 June 2004); and Waggoner, P.E. 1994. How Much Land Can Ten Billion People Spar for Nature? Council for Agricultural Science and Technology, Ames IA.

TABLE 9.2	Animal Products and Their Yield in kCal/m²/Year		
Milk	420	Beef	130
Eggs	200	Cheese	40
Chicken	190	Fish	2
Pork	190		

MATERIALS

No specialized materials required

TASKS

1. Interview another person and write the items they had for breakfast in the left column of the chart on the following page.

   a. How many calories would you estimate are eaten for the breakfast? (Remember 1 calorie is the same as 1 kCal.)

   b. Use your estimate of calories and Tables 9.1 and 9.2 to determine how many square meters might have been needed to grow their food.

   c. If 2000 calories are consumed daily, what portion of the average daily amount of nutrition would you estimate is provided by the breakfast?

2. Determine if the food represents high, moderate, or low degrees of processing (including packaging). Individually or as a class, devise (and describe) a formula to calculate the amount of energy required to produce food so that you can rate food on a 1 to 10 scale. Include distance from production to consumption, amount of processing, amount of packaging, and other variables you think are relevant. What information would you need to make this formula be more than an estimate?

3. Estimate nutritional value on the basis of high, moderate, and low—if the breakfast meets 1/3 of a person's daily need for nutrients consider it high in nutritional value. (Food such as a candy bar, despite its high calories, has low nutritional value. A bowl of cereal might have the same calories but is probably much better for you in terms of nutrition.)

4. How efficient was the person's breakfast in terms of the energy to prepare it and in terms of nutrients it provides?

5. How does it compare to what a poor person in a developing nation would typically eat? (Use your textbook or the Internet to help you.)

6. How does the breakfast compare to what a middle-class person in northern Europe would typically eat? (Use your textbook or the Internet to help you.)

7. Assume efficiency of the breakfast is a function of the land area to grow the food, the degree of processing, the energy expended in driving to get the food, the distance the food traveled, the market cost, and the nutritional value.

   a. What might a formula for rating efficiency look like?

   b. At what point does food become inefficient or wasteful?

   c. Rate each food item on a scale of 1 to 10 based on these factors and your judgment.

8. Compare results with others in the class. What food came the furthest? What traveled the shortest distance? What foods took the most energy and resources to prepare? What are your thoughts after doing this interview and comparing results?

Your name: _____ Date:_____ Person you interviewed: _____

Breakfast interview

Food item and estimated calories	Square meters to grow	Low processed, moderate, or highly processed	Miles driven to purchase this food	Distance food traveled from origin 0–200 or 200–1,000 k or > 1000 k	Market Cost in $	Nutrition Value	Efficiency on scale of 1–10
Totals							Avg =

Note: make a key or guide that provides your interpretation of the column headings and assumptions you made in completing the chart.

References

Akoroda, M.O. 1998. Comparative Output of Calories from Starchy Food Crops in Sub-Saharan Africa. Trop. Agric. 75:257–262.

Brewer, R. and M.T. McCann. 1982. Laboratory and Field Manual of Ecology. Saunders College Publishing, Philadelphia.

Smil, V. 2001. Enriching the Earth: Fritz Haber, Carl Bosch, and the Transformation of World Food. MIT Press, Cambridge, MA.

U.S. Department of Agriculture, National Agricultural Statistics Service. 2004. Agricultural Statistics: An Annual Report. Available at http://www.usda.gov/nass/pubs/agstats.htm (verified 14 June 2004).

Waggoner, P.E. 1994. How Much Land Can Ten Billion People Spar for Nature? Council for Agricultural Science and Technology, Ames IA. Available at http://www-formal. stanford.edu/jmc/nature/nature.html (verified 14 June 2004).

Human Survivorship Changes

OBJECTIVES

- Be able to describe how human mortality and survivorship have changed in the past 200 years
- Be able to collect population data and generate mortality and survivorship graphs

KEY CONCEPTS & TERMS

✓ Population
✓ Survivorship curve

INTRODUCTION

A key factor in environmental science is **population**, a group of individuals of the same species living in the same area. The study of changes in population size for any species involves two primary components: birth rate and death rate (it also includes immigration and emigration). If these two primary rates remain equal, the population size will generally remain the same.

Americans in the 21st century on average have a much longer life expectancy than they had in the 19th century. Among the factors that explain our longevity are better health care, improved nutrition, better working conditions, higher education, and new technology. If we are living longer, does that fuel population growth? According to Bodkin and Keller (2003) the life expectancy in ancient Rome was about 30 years.[1] In many countries, life expectancy has not improved much since then; it is only 39 in Botswana and some other African nations. However, in America average life expectancy is now around 77.

Human birth and death data are used to generate survivorship curves. A **survivorship curve** is a graphical representation of the likelihood that an individual will survive from birth to a particular age as shown in Figure 10.1. By comparing survivorship curves for different periods of time, you can identify historical trends in population demographics over the decades. Birth and death data can be obtained from gravestones, city/county records, and newspaper obituary notices. In this version of a commonly practiced demographics exercise you will collect data and make survivorship curves.

---

[1]There is some debate over the actual number of years and some researchers think Roman life expectancy was even lower. A good discussion can be found in Parkin, T.G. 1992. Demography and Roman Society. Baltimore: The Johns Hopkins Press.

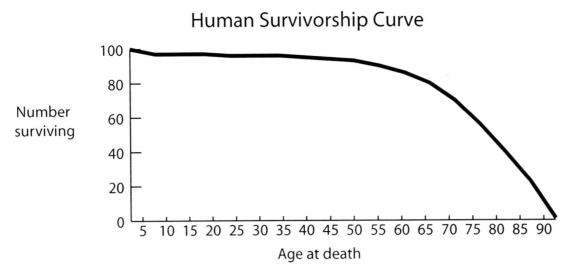

Figure 10.1—Sample Human Survivorship Curve.
After age 65, the number of survivors declines rapidly. If this population sample had experienced a major war or epidemic, the results might show up as one or more dips in the curve before the end of normal life expectancy.

MATERIALS

No special materials are required

TASK

Take a trip to a nearby cemetery to gather data on lifespan for a sample of people in the 19th century. You will also need to go to a library or use the Internet to gather data on a more current sample:

1. Examine headstones (or old newspaper obituaries) in a nearby community and record the age at death for males and females (25 each) who died prior to 1900. Do this during lab time. Each person is to gather his or her own data.

2. Record the age at death for males and females (25 each) in the community who died during the last five years. You can get this information from the obituary page in local newspapers, or on-line.

3. Enter the data in a spreadsheet (e.g., Microsoft Excel). If you are not already familiar with computer spreadsheets, this is your chance.

4. Create a graph from the data (use a computer spreadsheet to generate it unless your instructor tells you otherwise). You should have four lines on your graph, one each for pre-1900 females, pre-1900 males, contemporary males, and contemporary females. Put death age on the X-axis and number surviving on the Y-axis. This should be easy if you are using the data in spreadsheet format. Unless your instructor has told you otherwise, assume the graphs are to be computer generated. And remember—you are graphing survival curves not death curves. Use appropriate dimensions and labels for your graphs.

5. Once you have gathered the data and constructed the graphs, answer the following in a short (500 word maximum) typed essay to hand in along with the worksheets and graphs:

   a. Which group has the highest survivorship? (pre-1900 male, pre-1900 female, current male, Current female)?

   b. How large is the difference between highest and lowest survivorship?

   c. Why is there a difference? For this use information from your text and class notes <u>and</u> provide at least one current (no older than two years) peer-reviewed journal article citation to support your answer.

   d. What might the curves for the community look like in one hundred years? Support your reasoning.

CHAPTER

ELEVEN

Dowsing for Water

OBJECTIVES

- Be able to apply the scientific method
- Be able to think critically about scientific evidence

KEY CONCEPTS & TERMS

- ✓ Dowsing
- ✓ Energy audit

INTRODUCTION

Dowsing, otherwise known as "water witching," uses a "Y"-shaped diving rod or two L-shaped rods to search for underground water (or other resources). In recent years, it has also been used to search for buried water lines. How does dowsing work? Some theories suggest that there is a psychic connection established between the dowser and the sought object because all living things possess an energy force. By concentrating on the hidden object, the dowser is able to tune in to the energy force or "vibration" of the object, forcing the dowsing rod or stick to move. In this case, the diving tool may act as an amplifier or antenna for tuning into the energy. There is another view: dowsing is a myth based not on science, but superstition, and dowsers are successful in part because they "read" the landscape quite effectively and in part because groundwater is easily found in many places. Dowsing is controversial, and therefore presents an interesting example for the application of the scientific method.

Critical thinking is a key element in science. For this assignment, you will apply the scientific method to dowsing to determine some testable explanations.

MATERIALS

- ☐ Dowsing rod or stick

TASKS

There are at least five components or classes of variables involved in dowsing: the dowser, dowsing instrument, air, ground, and water. Can you think of some other categories or sources of variation?

With a partner, complete the following tasks:

1. Find an article or Internet site that is pro dowsing and an article or Internet site that is skeptical or critical of dowsing. Summarize what each document says. Does it seem credible? What strengths and flaws do you detect? Be sure to cite the sources as references.

2. Write a hypothesis for each of the five categories of variables. Remember that each of these five hypotheses must be testable.

3. Write a simple experimental design to test **one** of the hypotheses. Here is an example of how you could approach this: Suppose your hypothesis is that dowsing works for metal rods better than rods made from other materials because the metal rods "pick up" electromagnetic emanations from the water. After all, metal is a good conductor. You could test by controlling for everything except the type of material in the rod. Use different types of metals and also glass, plastic, or wood. In this experiment you might suggest blindfolding the testers so they do not know which type of rod they are using.

   a. In the above example, how many testers would you suggest and why?

   b. How many trials (repeats) would you suggest for each material?

   c. Now suggest your own experiment for a variable you choose to examine, and describe it.

4. Select one of the factors and conduct the experiment you described above.

   a. What were the results?

   b. Do the results support your hypothesis?

5. Did dowsing "work" for your subject?

6. Based on your results, would you say dowsing is based on a scientific or mythical phenomena? How confident are you in your results? Why?

7. What other experiment can you conduct to increase your confidence?

8. How did this activity help you to think about the application of scientific reasoning to the environmental sciences?

TWELVE

Aquatic Species Diversity

OBJECTIVES

- Be able to identify biodiversity terminology and concepts
- Be able to sample the invertebrate residents of a pond or stream and estimate biodiversity

KEY CONCEPTS & TERMS

✓ Biodiversity
✓ Ecological diversity
✓ Functional diversity
✓ Genetic diversity
✓ Sequential Comparison Index (SCI)
✓ Species diversity
✓ Species evenness
✓ Species richness

INTRODUCTION

Biodiversity is a popular term used by politicians, the media, the public, and environmentalists. What exactly does it mean? **Biodiversity** (or biological diversity) is defined as the different life forms (species) and life-sustaining processes that can best survive the variety of conditions found on earth. More specifically, biodiversity includes **genetic diversity** (variety in genetic makeup within a species), **species diversity** (variety among species in habitats), **ecological diversity** (variety of habitats such as forests, grasslands, streams), and **functional diversity** (biological and chemical processes or functions). Human actions, such as agriculture, elimination of species, pollution, deforestation, war, and settlement, can reduce biodiversity. This reduction in diversity has both known and unforeseen consequences as ecosystems react to these disturbances. Loss of biodiversity can increase the vulnerability of an ecosystem to further perturbations by reducing stability and functionality.

This lab's focus is species diversity. Natural ecosystems contain a wide variety of species, which interact to form stable and functional communities. As discussed above, the numbers and proportions of different species found in an ecosystem are referred to as **species diversity**. There are two components to measure this diversity: species richness and species evenness. **Species richness** is the number of different species. **Species evenness** is the relative proportion of each species. The **Sequential Comparison Index**

(SCI) is a simple method used to evaluate biodiversity and environmental quality in aquatic ecosystems (Cairns et al, 1968). The index ranges between 0.0 and 1.0. Although this index is only a general approximation, it is useful for comparisons if properly done. A high index (0.6 or more) suggests high biodiversity, high water quality, and dominance of species intolerant to pollution. A low index (0.3 or below) indicates low biodiversity, low water quality, and species with high tolerance of pollutants.

MATERIALS

☐ Boots	☐ Notebook
☐ Dip nets	☐ Squeeze bottles
☐ Forceps	☐ Trays (white, metallic, or light color)
☐ Local macroinvertebrate key	☐ Waders

TASKS

In this lab, we will gather and sort samples of invertebrates from a local pond or stream. Knowing something about species diversity within a habitat provides information about the complexity of the ecosystem. Complexity sometimes correlates with stability because a complex system can be more resilient to stress. Environmental scientists use indices of diversity in conjunction with rareness (the spatially restricted occurrence of a species) and habitat evaluation to make inferences on the state of an ecosystem. An excellent basic approach to understanding diversity is to use a simple **Sequential Comparison Index** (SCI). This approach allows you conduct a rough picture of species diversity without having to identify the specific species, which can be time consuming and difficult to the inexperienced. Researchers can determine the SCI for a site and then check it again to determine if biodiversity has changed.

- ∎ In teams, use the dip net to take a sample of water from the edge of the pond/stream. Employ the following steps in collecting macroinvertebrates with a dip net:
  - ● Choose a suitable habitat (leaf packs, snag areas, banks, and sediment) to collect samples.
  - ● If collecting from a leaf pack, drag the dip net through the leaf pack collecting leaves and debris.
  - ● To collect from a snag area or a bank, drag across the snag or bank with the net to collect debris.

- ∎ Dump the net's contents into a tray. Sort the debris from the specimens. Use a squeeze bottle filled with pond water to gently wash away debris from macroinvertebrates. Some aquatic insects build nests and cocoons, so be very careful in differentiating between debris and a macroinvertebrate.

- ∎ Try to align the specimens into rows in the same relative positions that they landed in the tray. Keep specimens wet.

- ∎ Once you have at least 12 organisms, begin in the upper left corner of the tray and count horizontally, scoring the number of adjacent organisms that are the same. Each group of one or more organisms is called a **run**. A run ends when you encounter a different type of organism. At the end of the row, continue into the next row as if there was no break (i.e., this is a continuous process). Thus, if the last organism in a row is the same organism type as the first one in the next row, it is the same run. If you are unable to get 12 individuals despite repeated sampling, move to another location or pool your results with other investigators.

■ Each individual organism will be noted with a letter. When you encounter a different species, this represents new run: assign it a new letter in sequence. For example: a,a,a,b,b,a,c,c,b,b,a,a = 12 individuals in 6 runs. Now, divide the number of runs by the number of individuals to obtain the Sequential Comparison Index (SCI = runs/individuals). In this example, 6/12 = 0.5, therefore SCI = 0.5. The SCI will range between 0–1. Low SCI number (0–0.3) very low diversity, whereas an SCI of 0.7–1 is high diversity. Figure 12.1 shows another example of how to determine an SCI.

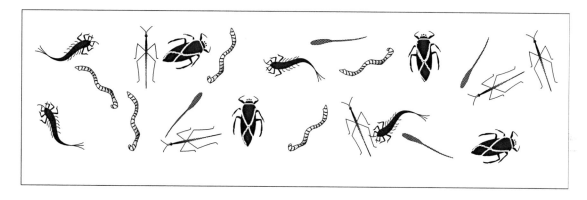

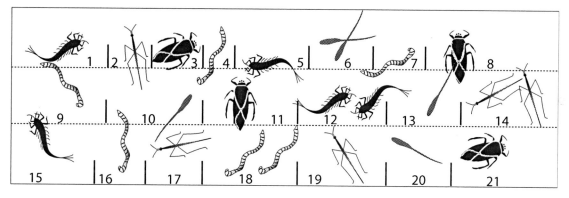

$$SCI = \frac{No. \, of \, Runs}{No. \, of \, Individuals} = \frac{21}{25} = 0.8 = \text{very diverse}$$

Figure 12.1—Determining a Sequential Comparison Index (SCI).
The SCI provides an easy way to obtain an approximation of species diversity and ecosystem health. Ideally, it should be used in conjunction with other tests.

■ Repeat this process for a total of five times. If you have a large enough sample, choose a different area in the pond/stream. If you do not get enough data to compare between your samples, sum up your trials into one sample and compare with the results of other students.

■ Sketch a map of the body of water and mark your sampling areas.

As individuals, answer the following questions:

1. For each sample site, what was the SCI?

2. Was there a difference in SCI between samples? Why do think there was or was not a difference?

3. What would you say about the species richness?

4. What would you say about the species evenness?

5. What factors (e.g., during collection) may have skewed the number and/or diversity of macroinvertebrates observed in your sample?

6. Were some macroinvertebrates more abundant in a particular area? Why?

7. What other ways could be used to determine species diversity? What could be done to calculate variance? What are some common measures used by scientists?

8. What conclusions can you make about the diversity of this pond/stream?

9. What other data would you want to collect to determine the ecological <u>and</u> environmental health of the pond/stream?

10. What human activities could adversely affect species diversity in this pond/stream?

11. How has this lab activity changed your thoughts about biodiversity in the pond or stream? Has it informed or changed your thoughts about biodiversity in general?

Reference

Cairns, J., D.W. Albaugh, F. Busey, and M.D. Chaney. 1968. The Sequential Comparison Index—A Simplified Method for Non-Biologists to Estimate Relative Differences in Biological Diversity in Stream Pollution Studies. Journal of the Water Pollution Control Federation 40:1607–1613.

Environmental Forensics

OBJECTIVES

- Be able to use several field and lab techniques to detect environmental contaminants
- Be able to report on potential sources of water contamination using an analytical and deductive approach

KEY CONCEPTS & TERMS

- ✓ Ammonia
- ✓ Biological oxygen demand
- ✓ Dissolved Oxygen
- ✓ Field screening techniques
- ✓ Laboratory analysis
- ✓ pH
- ✓ Point source discharge
- ✓ Specific conductance
- ✓ Turbidity

INTRODUCTION

In the environmental field, we often observe environmental conditions and problems without immediately knowing their cause. Some of these are chronic problems, such as frog deformations, species decline, and global climate change, and others are acute problems, including beaching of whales, viral infections of deer, and human poisonings. Environmental scientists are called into an investigation to determine what happened, how to remediate the problem, and how to prevent future problems. This backtracking, investigative approach is sometimes referred to as environmental forensics. Essentially, you are an environmental detective.

The approach is to use all available evidence to find the likely cause. Chemical and physical analyses are crucial, but are not the only methods to identify the root cause of a problem. Anecdotal evidence, experience, field observations, and previous studies and reports are all essential in solving the problem.

BACKGROUND

On August 26, a massive fish kill occurred in the Mountain River approximately 1 km downriver from Mountain City. Various local residents reported the kill to the state Fisheries and Wildlife Department on the morning of the 26th. Some residents claimed that they saw hundreds of rainbow trout (*Salmo gairdneri*) floating on their sides and bellies. According to one report, some of the trout were gasping. There were conflicting reports about visual water quality. Some reports indicated that the water was darker than normal. Also, some reports stated that there was some type of "foam" on the water. No oil sheens were reported. No reports of hazardous materials spills were made to the state police, local fire department, or the National Response Center.

A local resident recorded the river temperature, where the largest concentration of dead fish was observed, which was 26° C. The measurement was taken just below the surface. On August 26, the weather was reported as cloudy and overcast with an ambient temperature of 30.5° C. The 8 days prior to the 26th saw unusually high ambient temperatures, no precipitation, and uninterrupted overcast. At the nearby airport, the winds on the 26th were reported as relatively calm, 7 km/hr from the north.

The Mountain River watershed is approximately 125 km². Except for Mountain City, there is relatively sparse human population in the watershed. The housing stock within the river's watershed is primarily vacation homes and hunting camps. As shown in Figure 13.1, there are three facilities permitted by the state to discharge effluent into the river through point sources: Mountain City sewage treatment plant, Riverside Brewery, and ABC Chemical Company. (A **point source** discharge is a discrete, fixed discharge point

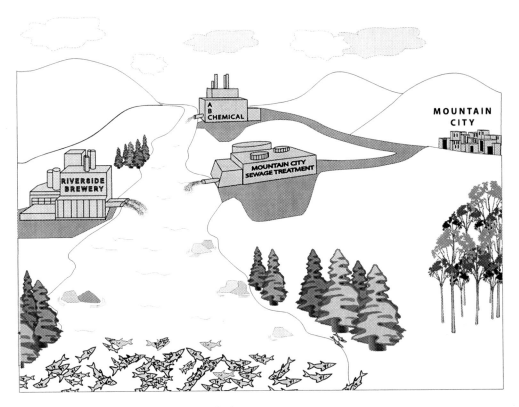

Figure 13.1—Mountain River Point Source Discharges.

This graphic represents the study area and the general location of the different discharges, which intermingle as the water flows downstream. (Not to scale.)

such as a pipe, culvert, ditch, or tunnel.) All three facilities are located in Mountain City, and range from 0.9 to 1.8 km upstream from the reported fish kill. Note that with point source discharges, because of the physical characteristics of rivers, there are varying zones of pollution impact as shown in Figure 13.2.

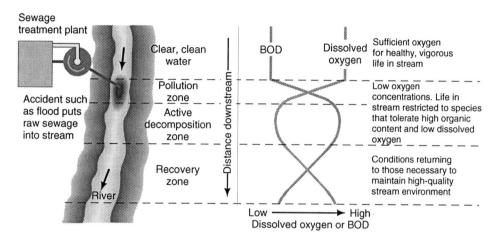

Figure 13.2—Sample Pollution Impact Zones in a River.
The input of sewage affects the relationship between dissolved oxygen and biochemical oxygen demand (BOD). Dead organic matter produces BOD, which can cause the amount of dissolved oxygen to become too low to sustain life. Source: Bodkin, D.B. and E. A. Keller. 2003. Environmental Science: Earth as a Living Planet. 4th ed. John Wiley & Sons, NY. (p. 425).

The Mountain City sewage treatment plant is relatively small. It is 26 years old, and serves approximately 250 residents and 8 industrial dischargers. The plant employs primary and secondary treatment and uses chlorine as a disinfectant. The Riverside Brewery is located across from the sewage treatment plant, but is not connected to the sewage plant and discharges directly into Mountain River. Approximately 0.9 km upstream from these two facilities is the ABC Chemical Company. This company does not manufacture chemicals, but purchases chemicals in bulk and repackages them for the consumer market, primarily automobile parts stores. Their primary products are sulfuric and phosphoric acids.

Approximately 25 years ago, the State Fish and Wildlife Department began stocking rainbow trout (S. *gairdneri*) in Mountain River. The fishery has become well established and is recognized for producing trophy trout.

In general, trout tend to be a more sensitive species, environmentally speaking. Like all species, S. *gairdneri* has a range of tolerance for environmental conditions (the concept of tolerance is explored in Lab 7; see Figure 7.1). Although they prefer cold, well-oxygenated waters, rainbow trout are capable of surviving in waters as high as 29.4°C (85°F) provided the water remains well aerated. Rainbow trout generally feed off the bottom foraging on virtually all aquatic insects and their larvae.

The following are common parameters of surface water quality and are potentially relevant to the fish kill:

■ Specific conductance
■ Dissolved oxygen
■ Turbidity
■ Ammonia
■ pH

SPECIFIC CONDUCTANCE

Conductivity is a measure of the ability of water to pass an electrical current and is an <u>indicator</u> of potential water pollution. Conductivity in water is affected by the presence of inorganic dissolved solids such as chloride, nitrate, sulfate, and phosphate anions (ions that carry a negative charge) or sodium, magnesium, calcium, iron, and aluminum cations (ions that carry a positive charge). Organic compounds like oil, phenol, alcohol, and sugar do not conduct electrical current very well and therefore have a low conductivity when in water (EPA, 2004a). Conductivity is also affected by temperature: the warmer the water, the higher the conductivity. Because conductivity readings can fluctuate with temperature, a standardized measurement called Specific Conductance (SC) is often used. For this reason, conductivity is reported as conductivity at 25°C and is an expression of a given sample's conductivity value standardized to 25°C, it can be calculated from known conductivity and temperature values as follows:

$$SC = \frac{conductivity}{1 + 0.0191\ (t\text{-}25)}$$
Where t = temperature of sample

The basic unit of measurement of conductivity is siemens.[12] Conductivity is measured in microsiemens per centimeter (μS/cm). Distilled water has a conductivity in the range of 0.5 to 3 μS/cm. The conductivity of rivers in the U.S. generally ranges from 50 to 1500 μS/cm. Studies of inland fresh waters indicate that streams supporting good mixed fisheries have a range between 150 and 500 μS/cm. Conductivity outside this range could indicate that the water is not suitable for certain species of fish or macroinvertebrates. Industrial waters can range as high as 10,000 μS/cm (US EPA, 2004).

DISSOLVED OXYGEN

Dissolved oxygen is a measure of the amount of oxygen freely available in water. It is commonly expressed as a concentration in terms of milligrams per liter (mg/L) or as a percent saturation, which is temperature dependent. Percent saturation is the percent of the potential capacity of the water to hold oxygen that is present. DO levels fluctuate seasonally and over a 24-hour period. They vary with water temperature and altitude. Cold water holds more oxygen than warm water and water holds less oxygen at higher altitudes (EPA, 2004). Thermal discharges, such as water used to cool machinery in a manufacturing plant or a power plant, raise the temperature of water and lower its oxygen content. Aquatic animals are most vulnerable to lowered DO levels in the early morning on hot summer days when stream flows are low, water temperatures are high, and aquatic plants have not been producing oxygen since sunset (US EPA, 2004).

[12]The SC instrument measures electrical resistance, that is, the extent something can *resist* an electrical current, which normally is reported in ohms. The unit of conductance was originally a "mho" (*ohm* spelled backwards). More recently, the term "**siemen**" is used in accordance with the terminology of the International System of Units. Both "mho" and "siemen" are sometimes used in water quality reports. One siemen is equal to one mho. Because SC in natural waters is usually much less than 1 siemen/cm, SC is usually reported in microsiemens (1/1,000,000 siemen) per centimeter, or μS/cm.

The DO for surface water ranges from 0 in extremely poor water conditions to a high of 15 mg/L (15 ppm) in 0 degree Celsius (freezing) water. The optimal DO level for *S. gairdneri* is 7–9 mg/L, and the acute lethal limit for trout is 3 mg/L (US EPA, 1986).

TURBIDITY

Turbidity is the measurement of water clarity and is related to the amount of suspended solids in the water. Suspended solids are variable, ranging from clay, silt, and plankton to industrial wastes and sewage. Turbid waters become warmer as suspended particles absorb heat from sunlight, causing DO to decrease. Suspended solids also can clog fish gills. Turbidity is measured in nephelometric turbidity units (NTUs). A normal range for turbidity in river water has not been established. Turbidity in drinking water should be less than one NTU. The U.S. EPA standard for drinking water is 0.5–1.0 NTU (US EPA, 1986).

AMMONIA

Sewage is the major source of anthropogenic ammonia (NH_3) in surface waters. Ammonia comes from the decomposition of urea in urine and decomposition of nitrogenous materials in sewage. Ammonia is toxic to fish and aquatic organisms, even in very low concentrations. When levels reach 0.06 mg/L, fish can suffer gill damage. When levels reach 0.2 mg/L, sensitive fish like trout and salmon begin to die (US EPA, 1986).

pH

pH measures the acidity or alkalinity of a solution (Figure 13.3). Acid conditions are usually caused when there is an excess of hydrogen ions (H^+) present. Alkaline conditions occur when there is an excess of hydroxyl ions (OH^-) present. At pH 7 (known as neutral), there is an exact balance between OH^- and H^+ ions; thus, it is neither acidic nor alkaline. As shown in Figure 13.3, the pH scale is logarithmic, which means that there is a 10× change in the number of H^+ ions (acid causing particles) or OH^- ions (alkaline-causing particles) present for each change of pH of 1 unit. Thus, pH 5 is ten times more acidic than pH 6 and 100 times more acidic than pH 7. pH affects many chemical and biological processes in the water. For example, different organisms flourish within different ranges of pH. The greatest variety of aquatic animals prefer pH at between 6.5 and 8.0. pH outside this range reduces the diversity in the stream because it stresses the physiological systems of most organisms and can reduce reproduction. In addition, low pH can allow toxic elements and compounds to become mobile and "available" for uptake by aquatic plants and animals resulting in secondary toxic effects (US EPA, 2004).

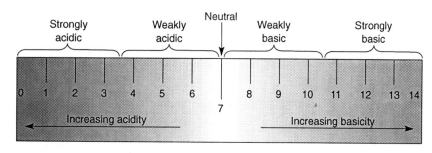

Figure 13.3—The pH Scale.
Pure water is neutral with a pH of 7. Each decrease of 1 pH unit means the acidity has increased tenfold.
Source: Raven, P.H. and L.R. Berg. 2004. Environment. 4th ed. John Wiley & Sons, NY (p. 479).

MATERIALS

There are four containers containing samples from: (a) upstream from Mountain City, (b) effluent from the Mountain City sewage treatment plant, (c) effluent from the Riverside Brewery, and (d) effluent from the ABC Chemical Company. Also on the tables, you will find various chemical, physical, and analytical testers, test kits, and reference materials:

- ☐ Ammonia test
- ☐ Conductivity meter
- ☐ Dissolved oxygen meter
- ☐ pH meter
- ☐ Thermometer
- ☐ Turbidity meter
- ☐ Other tests/meters

TASK

You will work in teams. The first step is to read the background material carefully and establish a plan of action. Your job is to test and analyze <u>every</u> aspect of the water with the provided equipment to identify the root cause of the fish kill. You will have to conduct some basic research to compare your findings with references.

Your lab report will be a memorandum to the director of the State Fish and Wildlife Department. Use the following format:

MEMORANDUM

TO:
FROM:
DATE:
SUBJECT:

Paragraph 1: (background, 2 to 3 sentences)

Paragraph 2: (what did you test for, summary of your results, uncertainties)

Paragraph 3: (what is your conclusion)

Paragraph 4: (what is your recommendation to the director)

Attachment: (a <u>table</u> presenting your analytical results)

NOTE: The results of your investigation are likely to be the basis of an enforcement action and will become public record. Therefore, it is imperative that your memorandum be written in neutral language and analysis of facts. Avoid inflammatory and value-laden terms and opinions.

References

U.S. Environmental Protection Agency (US EPA). 1986a. Quality Criteria for Water 1986, Washington, DC.

U.S. Environmental Protection Agency (US EPA). 2004. Monitoring and Assessing Water Quality [Online]. Available at http://www.epa.gov/owow/monitoring/ (verified 7 June 2004).

Field Sewage Treatment Plant

...unicipal or community wastewater

...issues in treating wastewater and

...t in the last quarter of the 20th century to ...1980s, the EPA provided more than $60 bil- ...and operated wastewater treatment projects ...tituted a significant contribution to the na- ...treatment plants, pumping stations, and the rehabilitation of aging sewer syst.... ...rogram led to the improvement of water quality in thousands of municipalities nationwide.

The purpose of a sewage treatment facility is to utilize bacteria to emulate and accelerate the natural process of human waste degradation. Increases in human population have increased waste flow to the extent that dilution can no longer be the solution to this pollution. The typical, modern sewage treatment plant employs primary and secondary treatment (Figure 14.1). Primary treatment removes debris and grit. Secondary treatment is used to break down the organic matter. To break down organic matter, microbes are used in combination with oxygen. Typically, three pounds of microbes are required for every pound of raw sewage in order to achieve acceptable levels of water clarity. Just before final discharge into a receiving water, the effluent is disinfected to reduce harmful bacteria using either chlorine gas, dry chlorine, ozone, or UV light.

However, sewage treatment plants are not restricted to sewage. Many area industries discharge their wastewater to the sewage treatment plant. Thus, the plant may also need to treat low levels of organic (e.g., solvents, petroleum) and inorganic (e.g., heavy metals, salts) pollutants.

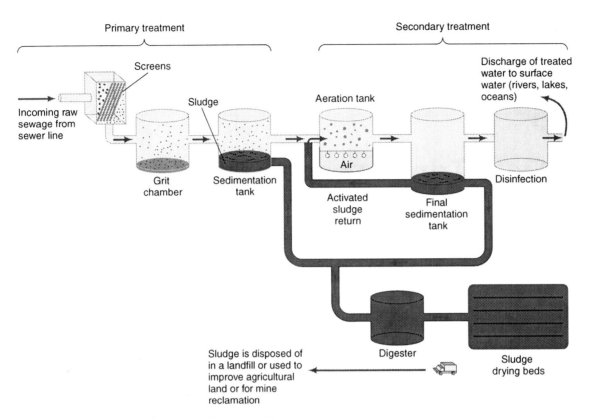

Figure 14.1—A Municipal Sewage Treatment Plant.

Source: Bodkin, D.B. and E. A. Keller. 2003. Environmental Science: Earth as a Living Planet. 4th ed. John Wiley & Sons, NY. (p. 440).

QUESTIONS TO ADDRESS IN RESPONSE TO THE FIELD TRIP

1. How does the plant work? How does it remove solids, reduce organic matter, and restore oxygen to the water?

2. What training do the operators and technicians have?

3. What types of parameters or conditions are monitored by the plant operators while the wastewater is being processed?

4. Who monitors the discharge of sludge (biosolids) to ensure that it meets standards? What are some of the standards? What happens if it the sludge fails to meet the standards?

5. How can sludge become a useful byproduct of the wastewater treatment system?

6. When is sludge considered a hazardous material?

7. Does the plant receive storm water (surface runoff) for treatment?

8. What are some special challenges of treating wastewater?

9. Is the water treated prior to its discharge? Is so, with what?

10. Where is the treated water discharged?

11. Who monitors the discharge of treated effluent to ensure that it meets standards? What are some of the standards? What happens when they are not met?

12. If water is safe to be discharged into the environment, why is it <u>not</u> used as the drinking water source for the community?

13. What seasonal changes (if any) are necessary in the operation of the plant?

14. What is the biggest water quality concern facing the plant? What plans, if any, do the operators have for reducing the concern?

15. As water users and waste generators, we all contribute to the water waste stream. (Even if you have an on-site private septic system in the service area, some waste eventually comes to the plant.) How should we decide what parts of the plant to upgrade, or how much money to invest in the plant?

Reference

U.S. Environmental Protection Agency (US EPA). 1998. Wastewater Primer (EPA 833-K-98-001). Washington, DC.

Field Trip: Wetlands Mitigation

OBJECTIVES

- Be able to describe the basic functions and importance of wetlands and the challenges faced in their protection
- Be able to report on the use and successes of various wetland mitigation efforts

INTRODUCTION

For this lab you will visit a wetland that has been enhanced or planned to compensate for on-site or adjacent wetland impacts. Or this wetland might be a site that has been evaluated to assess how it will stand up to potential impacts. Be prepared to ask questions during the tour. After the tour, write up a response incorporating what you have learned and addressing the questions below.

In the intervening centuries since the United States became a country, over half of our wetland areas have been lost or converted into other uses. Activities resulting in wetlands loss and degradation include: agricultural runoff and conversion to agricultural land; commercial and residential development; road construction; impoundment; resource extraction; industrial processes and waste; dredging; silviculture; and insect pest control (US EPA 1994; US EPA 1993). Pollutants from sediment, nutrient loading, pesticides, salinity, heavy metals, weeds, low dissolved oxygen, pH change, and selenium can degrade a wetland (US EPA 1994). Yet wetlands play a vital role in ecosystem health and environmental quality.

The U.S. government allows "mitigation banking" to offset the impacts of construction on federal lands, using federal funds, or through federal permits and licenses (US EPA 1995). A wetlands mitigation bank refers to one or more wetland areas that have been restored, created, enhanced, or (in exceptional circumstances) preserved, and are set aside to compensate for the loss of future wetlands due to development activities. A wetland bank may be created when a government agency, a corporation, or a nonprofit organization undertakes such activities under a formal agreement with a regulatory agency. The value of a bank is determined by quantifying the wetland values restored or created in terms of "credits." Sometimes the bank is immediately used and a mitigation area is proposed as part of the original project. Other times, the bank is held in reserve for future activities.

Wetlands mitigation has become fairly well established, although there are some concerns such as whether or not one wetland can truly replace another, and the ethics of wetlands removal in the first place. The U.S. Army Corps of Engineers (2001) has specific criteria for reviewing wetlands mitigation. Hammer (1997), Harker et al. (1999), Marble (1992), Mitsch & Jørgensen (2004), Mitsch & Gosselink (2000), Payne (1992), Thunhorst (1993), and many other sources exist for use in planning wetland mitigation.

FACTORS TO CONSIDER IN PROPOSING OR EVALUATING A MITIGATION SITE

- Are the proposed boundaries and characteristics of the compensation site documented, including elevation, sources of water, and proposed vegetation?
- Look for a narrative describing the specific goals of the compensation work in terms of particular wetland functions and values as related to those of the original wetland. (See Figure 15.1.) This narrative must also identify the criteria to measure success of the compensation work (e.g., water level within tolerances as defined in the proposal, percent survival of plants, etc.). If a narrative has not already been prepared, suggest some specific things to be addressed in the narrative based upon your site visit.

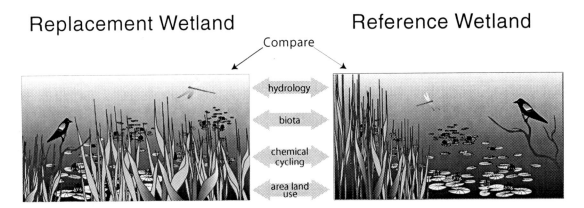

Figure 15.1—Comparison of Reference and Replacement Wetlands.
The U.S. has lost approximately half of its wetlands in the past 200 years (US EPA 2004). Replacement wetlands are one strategy to offset wetlands loss from the effects of development. The replacement wetland size and characteristics are determined by policy and on a case-by-case basis. The goal is to obtain a replacement wetland with similar functions and values and at least of similar benefit. A reference wetland is used for comparison; the replacement wetland, whether natural, artificial, or enhanced, should be ecologically self-sustaining and protected from development. Other considerations in wetland replacement policies include planting schemes, treatment of invasive species, hydrology, primary productivity, energy pathways, restoration/alterations plans, monitoring, evaluation, escrow and maintenance accounts, and wetlands banking policies.

- There should be a narrative describing the available literature or experience to date (if any) for carrying out the compensation work.
- Any evaluation of the site itself would not be complete unless it also addressed the context: social and political issues such as neighborhood values and uses of the sites, legal procedures for site approval, public relations issues, and construction concerns.

QUESTIONS TO ADDRESS IN RESPONSE TO THIS FIELD TRIP

1. What laws or rules define wetlands in your state? How do the state and local regulations mesh with federal regulations?

2. What agencies and environmental groups are involved in protection of wetlands? Which ones are involved in this particular wetland?

3. Is this site a new mitigation wetland? If so, how long will it take for this particular created wetland to meet the definition of a functional wetland based on soil, plant, and hydrology criteria? If not, describe this wetland in terms of current and eventual soil, plants, and hydrology.

4. If this wetland were to be used for mitigation, what could it mitigate, assuming a three to one ratio? Address size, characteristics, and location.

5. Compare this wetland to the nearest accessible wetland. How similar are they in terms of watersheds, wildlife corridors, aesthetics, hydraulics, and other functions?

6. What is the monitoring schedule for the wetland, and what are the determinants for "success"?

7. What are the key issues in wetlands compensation?

8. What do you think about the idea of enhancing wetlands or creating new wetlands as compensation for land development that damages or eliminates wetlands?

References

Hammer, D.A. 1997. Creating Freshwater Wetlands. 2nd. Lewis Publishers, Boca Raton.

Harker, D.G., L.K. Harker, S. Evans, and M. Evans. 1999. Landscape Restoration Handbook. 2nd ed. Lewis Publishers, Boca Raton.

Marble, A.D. 1992. A Guide to Wetland Functional Design. Lewis Publishers, Boca Raton.

Mitsch, W.J. and J.G. Gosselink. 2000. Wetlands. 3rd ed. Wiley & Sons, New York.

Mitsch, W.J. and S.E. Jørgensen. 2004. Ecological Engineering and Ecosystem Restoration. John Wiley & Sons, New York.

Payne, Neil. 1992. Techniques for Wildlife Habitat Management of Wetlands. McGraw-Hill: New York.

Thunhorst, G.A. 1993. Wetland Planting Guide for the Northeastern United States. Environmental Concerns, Inc., Maryland.

U.S. Army Corps of Engineers. 2001. Checklist for Review of Mitigation Plan (Draft 02/21/2001). United States Army Corps of Engineers, New England District Regulatory Branch.

U.S. Environmental Protection Agency (US EPA). 1993. Guidance Specifying Management Measures for Sources of Nonpoint Pollution in Coastal Waters. EPA: Washington, DC.

U.S. Environmental Protection Agency (US EPA). 1994. National Water Quality Inventory. 1992 Report to Congress. EPA 841-R-94-001. Washington, DC.

U.S. Environmental Protection Agency (US EPA). 1995. Wetlands Fact Sheets [Online]. Available at http://www.epa.gov/OWOW/wetlands/facts/fact16.html (verified 9 June 2004).

U.S. Environmental Protection Agency (US EPA). 2004. Wetlands [Online]. Available at http://www.epa.gov/owow/wetlands/vital/status.html (verified 9 June 2004).

Field Trip: Water Treatment Plant

OBJECTIVES

- Be able to describe the process by which a water district obtains, treats, and distributes drinking water
- Be able to describe key issues associated with maintaining a drinking water system

INTRODUCTION

Water districts supply water to large areas and may also treat their sewage wastes (Figure 16.1). If you have a public water supply, the district treats the water to destroy possible disease-causing microorganisms like bacteria and protozoa. Ozone treatment is an alternative to the intense chemical and filtering processes that occur in most water treatment facilities.

This field trip explores where your drinking water originates, how it is treated and distributed, and how the resource is protected. Obviously, the cleaner the water source, the less treatment it requires. Thus we should all be concerned about the activities occurring in our watershed because it is the source of our potable water.

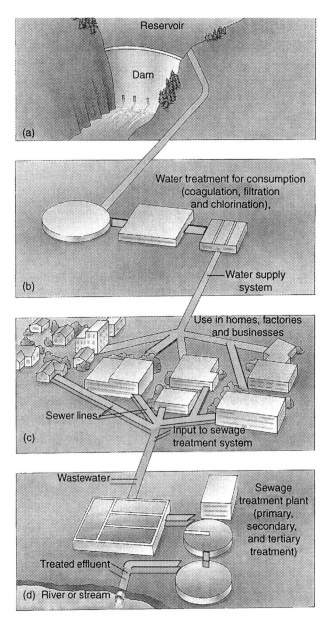

Figure 16.1—A Municipal Water Treatment System.
(a) A water supply may come from a groundwater reservoir (as shown above). (b) Treatment begins with co-agulation to clump suspended particles, filtration to remove the clumps and ozone, chlorination, or other treatment to destroy microorganisms. Then the water is distributed for use. Residual chlorine in the pipes helps keep it safe. (c) The wastewater is collected by municipal lines. (d) Primary, secondary, and occasionally tertiary treatment prepares the effluent for discharge into "receiving waters." Source: Raven, P.H. and L.R. Berg. 2004. Environment. 4th ed. John Wiley & Sons, NY (p. 502).

QUESTIONS TO ADDRESS IN RESPONSE TO THIS FIELD TRIP

1. What is the source of the plant's water?

2. How much water is needed by the communities served? How much water per person?

3. Is the demand for water expected to change as a result of population change or other reasons? If so, what are the plans to address this increased demand?

4. What are the plans to reduce water demands in the event of a short-term emergency?

5. What are the plans to deal with long-term problems such as drought?

6. How much and what type of treatment is employed to make the water potable?

7. What is the difference between the ozone used in ozone purification treatment compared to the ozone layer (stratosphere) around the Earth?

8. What are the energy needs to treat water? What are the by-products? What would happen in the event of a power loss?

9. What is "residual chlorine," and why is it important?

10. Are there major microorganism pests of concern with the water? If so, what are they?

11. What are the biggest water quality concerns facing the watershed?

12. Does the plant provide environmental education efforts appropriate for this operation? Comment on the need for education concerning water treatment.

CHAPTER

SEVENTEEN

Air Quality and Automobiles

OBJECTIVES

- Be able to describe the difference between samples and populations
- Be able to locate air quality information from reputable Internet sites
- Be able to summarize the major air quality impacts of the automobile
- Be able to estimate the average amounts of annual pollutants emitted by cars at a university or business

MATERIALS

- ☐ Computer access (Internet and Excel)
- ☐ Clipboard
- ☐ White plastic squares (about 10 cm × 10 cm)
- ☐ Strong tape (packing or duct tape)
- ☐ Petroleum jelly
- ☐ Magnifying lens (or dissecting microscope)

INTRODUCTION

The automobile is a major source of air pollution. In 1998, nationwide, personal vehicles (trucks, cars, and SUVs) were responsible for approximately 31% of nitrogen oxides emissions, 29% of volatile organic compound emissions, 21% of particulate matter, and 57% of carbon monoxide emissions (US EPA, 2000). The automobile also is a major source of carbon dioxide, a greenhouse gas. As discussed below, these pollutants present potentially significant health and environmental risks. In addition, as depicted in Figure 17.1, the automobile contributes a number of secondary pollutants. Although emissions from an *individual* car are generally low, in numerous cities across the country, the personal automobile is the single greatest polluter, as emissions from millions of vehicles add up. Driving a private car is probably a typical citizen's most "polluting" daily activity and is the most significant contribution to environmental degradation (US EPA, 1994).

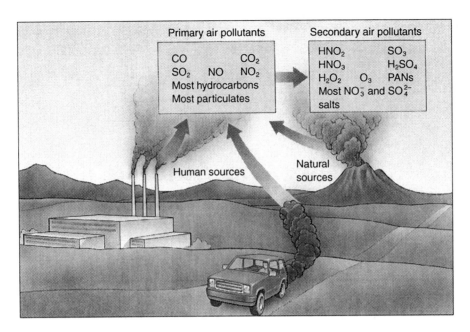

Figure 17.1—Air Pollutants and the Automobile.

The automobile is a major source of air pollution. Primary pollution refers to the emitted or discharged pollution, when it reacts in the atmosphere. Secondary pollutants are produced from chemical reactions involving the primary pollutants. Source: Raven, P.H. and L.R. Berg. 2004. Environment. 4th ed. John Wiley & Sons, NY (p. 439).

HYDROCARBONS

Hydrocarbon emissions (or volatile organic compounds, VOCs) occur when fuel molecules in the internal combustion engine do not burn completely. Hydrocarbons react in the presence of nitrogen oxides and sunlight to form ground-level ozone, a major component of smog. Ozone irritates the eyes, damages the lungs, and aggravates respiratory problems. It is the nation's most widespread and intractable urban air pollution problem (US EPA, 2002). A number of exhaust hydrocarbons are also toxic and carcinogenic. Hydrocarbon pollutants also escape into the air through fuel evaporation. Evaporation occurs primarily during refueling, while the engine is running, and right after the engine is shutdown (US EPA, 1994).

NITROGEN OXIDES

Under the high pressure and temperature conditions in an engine, nitrogen and oxygen atoms in the air react to form various nitrogen oxides, collectively known as NOx. Nitrogen oxides, like hydrocarbons, are precursors to the formation of ozone. Nitrogen oxides also contribute to the formation of acid rain (US EPA, 2000).

PARTICULATE MATTER

Particulate matter refers to solid or liquid particles found in the air. Some particles are large or dark enough to be seen as soot or smoke, but fine particulate matter is tiny and generally not visible to the naked eye. Particulate emissions from automobiles generally consist of very tiny particles, also known as PM2.5, because they are less than 2.5 microns in diameter. It is a health concern because very fine particles can reach and lodge in the deepest regions of the lungs. Health effects include asthma, difficult breathing, and chronic bronchitis, especially in children and the elderly. Fine particulate matter associated with diesel exhaust is a suspected carcinogen. Fine particulate matter can travel long distances and is a major cause of haze, which reduces visibility, affecting cities and scenic areas (US EPA, 2000).

CARBON MONOXIDE

Carbon monoxide (CO) is a product of incomplete combustion and occurs when carbon in the fuel is partially oxidized rather than fully oxidized to carbon dioxide (CO_2). Carbon monoxide reduces the flow of oxygen in the bloodstream and is particularly dangerous to persons with heart disease (US EPA, 2000).

CARBON DIOXIDE

Carbon dioxide (CO_2) is a product of complete combustion. Although carbon dioxide does not directly impair human health, it is a "greenhouse gas" that traps the Earth's heat and contributes to global warming (US EPA, 2000).

TASKS

In this lab, we will be conducting some basic measurements and calculations to estimate the area impact of vehicles used by your university community (students, staff, and faculty) and the contribution of vehicle-based pollution from your county, your state, and the nation. To determine this, you will be estimating the annual average emissions produced by an average car at the university.

I—PARTICULATE MATTER ASSESSMENT

- Coat one half of each of the two white plastic squares with a thin, even coat of petroleum jelly. Cover the other half with tape. (The tape is also used to fasten the squares, jelly side up, to a platform.)
- Place one square outdoors near a parking lot in a secure area so that it will not be disturbed. Place the other square inside the lab.
- After 24 hours, retrieve each plastic square and carefully remove it from the platform.
- Using the magnifying glass or dissecting microscope, count the number of particulate matter particles on the samples.

1. Compare the results from each sample. Describe what you see. Which areas contained the most particulates?

II—PARKING LOT ASSESSMENT

2. As teams, you will be assigned a parking lot. At each parking lot, count and record each car and the make, model, and year (e.g., Honda, Civic, 2003). If you do not know the exact year, make a reasonable guess as to the age. (This raw data must be submitted with your report.) Be sure to record the number for each category if there are multiple cars. (Use the attached Parking Lot Tally Sheet.)

TABLE 17.1	Vehicle Tally Example	
Make	**Model**	**Year**
Honda	Civic	2003
VW	Golf	1990
Ford	Explorer	1999

3. Using a computer and Internet access, go to the Environmental Defense Fund's Tailpipe Tally site http://www.environmentaldefense.org/tailpipetally. Input each automobile (assume 12,500 miles per year) and record the following information (fuel consumption, CO_2, CO, NO_x, and HC):

TABLE 17.2	Vehicle Emission Tally Example				
Car (assumes) 12,500 miles per year	**Annual fuel consumption (gallons)**	**Annual carbon dioxide (CO_2)**	**Annual carbon monoxide (CO)**	**Annual nitrogen oxides (NO_x)**	**Annual hydrocarbons (HC)**
Honda, Civic (SULEV) 2003	262	5088	79.0	0.8	0.8
VW, Golf 1990	481	9327	344.9	44.9	25.9
Ford, Explorer, 4WD 1999	750	14540	399	44.1	21.2
TOTAL	1,493 gal	28,955 lbs	822.9 lbs	89.8 lbs	47.9 lbs

4. After completing this chart, input the data into a Microsoft Excel spreadsheet. As a team, you will combine all the data using Excel to calculate a parking lot sample mean for: fuel consumption, CO_2, CO, NO_x, and HC. (The Excel spreadsheet must be submitted with your write-up.)

5. Using the data obtained in step 4, use the <u>mean</u> for each pollutant category and fuel consumption to calculate the total amount, per pollutant category, and fuel consumption, contributed by:

   a. The university community;[13]

   b. All the registered vehicles in your city/county (start by checking your state or county motor vehicle department Internet sites); and

   c. All the registered vehicles in your state

   NOTE: Remember, you almost never have the exact data you need. So, you will have to make assumptions and/or extrapolations. Clearly identity what assumptions you made and briefly explain why you chose a particular assumption.

---

[13]You may have to consult your university department of public safety to identify how many parking permits are issued by group (faculty/staff and students).

6. Now, use this same data, but for the country. Using the *Statistical Abstract of the* U.S. (http://www.census.gov/prod/www/statistical-abstract-us.html), find the total number of cars registered in the U.S. (Hint: go to *Transportation*, Table 1084.) Using your data, calculate the national, total amount of pollutants contributed in each pollutant category and the total fuel consumed.

7. What about your contribution to pollution?

   a. What are the make, model, and age of <u>your</u> car? (If you do not own a car, use your parents' or the lab instructor's car.)

   b. How much fuel does it consume and how much pollution does it generate?

   c. Assume everyone at the university, in your city/county, in your state, and in the U.S. had your car. Calculate the total annual amount of fuel consumed and the amount of <u>each</u> air pollutant generated for the:

      i. University

      ii. City/county

      iii. State

      iv. U.S.

   d. What are the percent differences between your car and the parking lot means you used to answer Questions 3 and 4?

WRITE-UP

For this lab, a formal lab write-up is required:

Title
Introduction—What did you do and why did you do it. Also, state your hypothesis (e.g., your car compared to the sample mean).
Materials & Methods—How did you do it?
Results—Your tables
Discussion—In your discussion:

1. Describe potential areas of weaknesses in your conclusions (e.g., sampling, extrapolation, assumptions).

2. Briefly describe what you would do next time to improve the results.

3. Examine and reflect on this data as an individual. That is, what is the environmental significance of your findings? Were you surprised by any of the car comparisons in your data? (For example, note that the Ford Explorer in Table 17.2 uses much more gasoline than the VW Golf but actually produces fewer nitrous oxides and hydrocarbons.)

4. Briefly describe some appropriate policy recommendations to reduce university automobile pollution.

TABLE 17.3	Parking Lot Tally Sheet	
Make	**Model**	**Year**

References

U.S. Environmental Protection Agency (US EPA). 1994. Automobile Emissions: An Overview (EPA 400-F-92-007). GPO, Washington, DC.

U.S. Environmental Protection Agency (US EPA). 2000. National Air Pollutant Emission Trends: 1900–1998 (EPA 454/R-00-002). GPO, Washington, DC.

U.S. Environmental Protection Agency (US EPA). 2002. Latest Findings on National Air Quality: 2001 Status and Trends (EPA 454/K-02-001). GPO, Washington, DC.

EIGHTEEN

Indoor Air Quality

OBJECTIVES

- Be able to describe the major sources of indoor air pollution
- Be able to measure carbon dioxide levels, humidity, temperature, and flow in the air
- Be able to describe how indoor air pollution problems are reported and handled

KEY CONCEPTS & TERMS

- ✓ Air exchange rate
- ✓ Indoor air quality
- ✓ Infiltration
- ✓ Mechanical ventilation
- ✓ Natural ventilation
- ✓ Sick building syndrome

INTRODUCTION

Air pollution has been around for a very long time. In 4000 B.C. people forged copper and gold and baked and glazed pottery creating air pollution. Civilization began burning coal and coke for energy in 1000 A.D., and by 1306 A.D. coal burning was prohibited in London due to the extensive smoke and smog pollution (Jacobson, 2002). (Passing a law is one thing, enforcing it is a different story.) In 1963, the U.S. Congress was concerned enough about air pollution that it passed the first Clean Air Act, but this first law was ineffectual. The Act was extensively amended in 1970 and 1990 to regulate mobile and stationary sources of air pollution, decrease acid rain, and protect stratospheric ozone.

In the early 1970s the oil shortage prompted a national focus to conserve energy. One outcome was to reduce the loss of heat in buildings due to poor insulation and ill-fitting windows. At the same time, industry was increasingly manufacturing household products and building materials using plastics, solvents, glues, and other chemicals (e.g., form-aldehyde) as shown in Figure 18.1. This presents a potential exposure problem as most Americans spend up to 90% of their time indoors and many spend most of their working hours in an office environment. Studies conducted by the U.S. EPA and others show that indoor environments sometimes can have levels of pollutants that are actually higher than levels found outside (US EPA, 1997).

As flow of fresh air was being reduced into buildings, pollutants were being added, including second-hand tobacco smoke, which affected **indoor air quality**—the nature of air that affects the health and well-being of occupants. This resulted in the rise of **sick building syndrome**, defined as a set of symptoms that affect some occupants during time spent in a building and diminish or go away during periods away from the building, but cannot be traced to specific pollutants or sources within the building.

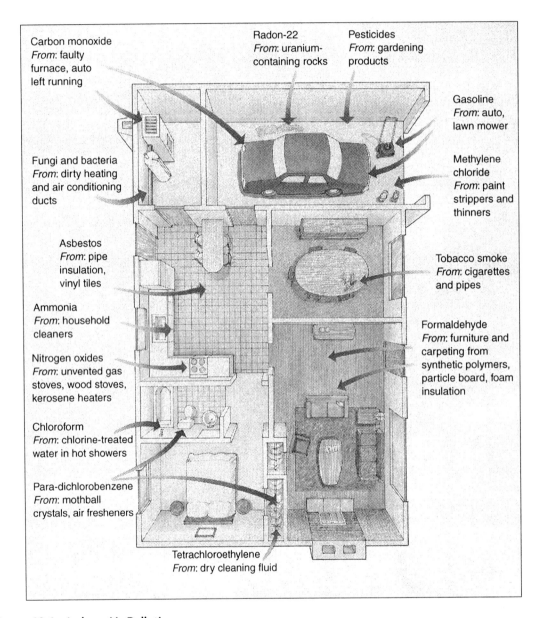

Figure 18.1—Indoor Air Pollution.
Air quality inside a building cannot be any better than the air outside, but good air ventilation can make a big difference. Source: Raven, P.H. and L.R. Berg. 2004. Environment. 4th ed. John Wiley & Sons, NY (p. 454).

Ambient (outside) air enters and leaves a structure through infiltration, natural ventilation, and mechanical ventilation. With **infiltration**, outdoor air flows in through openings, joints, and cracks in walls, floors, and ceilings, and around windows and doors. In **natural ventilation**, air moves through opened windows, doors, and passive vents. **Mechanical ventilation** devices such as outdoor-vented fans are used to intermittently remove air from a room, to air handling systems that use also distribute filtered and conditioned outdoor air to strategic points throughout the house. The rate at which outdoor air replaces indoor air is described as the **air exchange rate**. When there is little in-filtration, natural ventilation, or mechanical ventilation, the air exchange rate is low, and pollutant levels will build up.

An indoor air quality problem exists when four elements are present: a contaminant, a pathway to a person, a driving force to move the contaminant, and a person exposed to the contaminant. If any of these elements is removed, the problem is eliminated. For ex-ample, a chemical leak in an unoccupied warehouse is not a direct air quality problem (no exposure means no human health effect). Although there is a source, a pathway (the air in the building), a driving force (drafts and ventilation that circulate air), if the building is unoccupied there will be no exposure. In comparison, if there is mold contamination in ductwork of an occupied office building, there is a source of contamination (mold), a pathway (the duct work), a driving force (the air through the duct), and human exposure (occupants of the building).

To understand the complex issues involved with indoor air quality you need to recog-nize potential contaminants, understand types and symptoms of exposure, health effects of the contaminants, understand properties of air and heating, ventilation and air condi-tioning (HVAC) systems, and determine how to evaluate exposures. In this lab we will in-vestigate indoor air quality using simple, common techniques.

MATERIALS

- ☐ Calculator
- ☐ CO_2 meter
- ☐ Bottle of soap bubble solution
- ☐ Digital thermometer with relative humidity readout

TASKS

Task 1

For this task, read the information below and then conduct an investigation into the in-door air quality of a building. The questions below will guide you in writing your report. The air we breathe is actually a combination of many gases. The major components of air in percentage of volume are:

Nitrogen	78.08%
Oxygen	20.95%
Argon	0.93%
Carbon dioxide	0.03%

In addition, air contains particulate matter, volatile organic compounds, "biologicals," and water vapor. Air has a weight density of 75 lbs per 1,000 cubic feet. Water, by com-parison, weighs 830 times as much as air. Air has a composite molecular weight (MW) of about 29. The amount of moisture in air is limited by temperature and pressure. The rel-ative humidity is the percentage of water in the air compared to the maximum amount of water the air will hold at that temperature.

Air's density (weight to volume ratio) varies with temperature and pressure according to the Ideal Gas Law:

$$PV = nRT$$

P = pressure, V = volume, nR = constant, and T = temperature

Standard temperature and pressure (STP) is assumed to be 20° C and 760 mm Hg. Correction factors are used when air is measured at other than standard conditions to account for the change in air density.

Pressure differences cause air to move. Therefore, for a contaminant to move from a source to a person there must be a change in pressure. In the atmosphere, pressure differences create movement of large air masses, which we call wind. Atmospheric pressure pushes air from high pressure to low pressure. In buildings air is moved in ducts and a fan is used to create a pressure difference. A fan lowers atmospheric pressure (creates negative pressure), and atmospheric pressure moves air into the duct in order to equalize the pressure. Negative pressure wants to collapse the duct, positive pressure wants to blow it up. Pressures are used in buildings to control temperature, airflow, and contaminants. For example, reactor containment buildings at nuclear power plants are under negative pressure; a small radioactive leak would tend to stay inside the containment building. To ensure this condition, less air is supplied to the containment building than to the surrounding environments. Air from the surrounding areas will be pulled into the building to compensate for the negative pressure differential, thereby eliminating the possibility that radioactive gases can leak out.

While sophisticated equipment exists to measure exact pressures in ducts and building supply and exhaust vents, we will use some very simple techniques to determine pressure differentials in a building. Imagine you are being asked to write a report in which you investigate the air quality of a building. You will be testing:

- Carbon dioxide levels
- Pressure and ventilation as indicated by airflow
- Temperature
- Humidity
- Signs of potential biological contamination (e.g., mold, water stains, moisture buildup, and odors)

Be discrete and safe in looking around so that you do not place yourself at risk.

Soap bubbles can help you to see airflow and determine the pressure of spaces in the building. Remember that air moves from positive pressure to negative pressure.

1. Fill out Table 18.1 as you investigate the pressure relationships in the building. Incorporate this information into your report.

2. Try releasing some soap bubbles near a supply or exhaust vent in the room. Describe what happens.

3. Release soap bubbles near the door to the lab or classroom to determine if the room is under positive or negative pressure. Select other places in the building to help you determine air pressure patterns.

TABLE 18.1 Room Pressures

Location	Air Movement Direction	Pressure (+ or −)

There are many ways to evaluate air quality and many methods to collect air samples, measure gases, vapors, and particulates and analyze for biological exposures. These methods are complicated and depend on knowledge of specific chemicals and concentrations. Many air quality complaints are associated with environmental comfort issues such as temperature, relative humidity, and air velocity. A simple first step of an indoor air quality investigation is to measure temperature and relative humidity of the room. These levels can be compared to recommended guidelines for temperature, humidity, and airflow for summer and winter conditions.

The most common connection to poor air quality is inadequate ventilation. The movement of the soap bubbles gives you some idea about flow. Another approach is to look at carbon dioxide (CO_2) levels: if an adequate quantity of air is provided to the occupants of the building, then the CO_2 levels are reasonable.

Ventilation systems are designed to support a maximum occupancy of a space. Sometimes a room is fully occupied, other times minimally occupied, and still other times it may even be above capacity. Rather than review ventilation plans and measure the amount of air entering the building and the amount reaching the occupants, an "indicator" can be used. When we breathe we inhale oxygen and exhale CO_2. If the building ventilation system is supplying and exhausting an appropriate amount of air to a space, levels of exhaled CO_2 will not build up in a room. (Sustained levels of CO_2 above 1000 parts per million (ppm) usually indicate inadequate ventilation.) Accordingly, CO_2 can be used as an indicator of appropriate ventilation. During an initial indoor air quality investigation, CO_2 levels are measured throughout the day and analyzed to determine if the ventilation system is working correctly.

4. As part of your investigation, determine good locations to test the flow of air into and from the building and the spaces within. Use Table 18.2 (enlarge as needed). Be sure to also sample the ambient air—the outside of the building. Since you are looking at problem spots, be sure to select confined spaces—phone booth, bathroom, closet, elevator—for CO_2 levels, soap bubble test, temperature, humidity, and biological inspection.

TABLE 18.2 Instrument Evaluation

Location	Temperature	Relative Humidity	CO_2	Biological Inspection
Ambient				

5. Compare your results to the American Society of Heating, Refrigeration and Air Conditioning Engineers (ASHRAE) guidelines or a federal agency's set of guidelines for temperature and humidity. Your instructor may supply you with these guidelines or ask you to research them on the Internet or in the library. How does your building (and individual rooms) compare? Are they within the range for the season?

6. Is the building providing adequate ventilation? Why or why not?

7. What can you say about the connection between ambient air quality and indoor air quality from these results?

8. What recommendations can you make to ensure good air quality in this building, based upon your results and considering the need to conserve energy?

9. Evaluate your air quality testing strategy. What would you change if you had more time and resources?

Task 2

Answer the following questions:

10. How are ambient air and indoor air quality connected?

11. What contaminants are found in outdoor air? What are the major sources?

12. What contaminants are found in indoor air? What are the major sources?

13. Indoor air quality is a major public health concern because the average person spends the majority of their time indoors, potentially exposed to many indoor contaminants. Keep a log tracking your indoor time for one day. Construct a table (as per example, Table 18.3) with four columns to list the (a) location, (b) amount of time, (c) activity and (d) potential contaminants to which you are exposed.

Calculate the percentage of time you spent indoors.

$$(T_{in\ (hours)} \div 24\ hrs) \times 100 = \%\ Indoors$$

$$T_{in} = Total\ time\ indoors\ (hrs)$$

TABLE 18.3   **Indoor Activity Log**

Location	Time Spent	Activity[14]	Potential Contaminants

[14]Activity is important because as activity increases, the quicker and deeper the breathing (e.g., sleep versus exercise) and thus, increased exposure to potential contaminants.

References

Jacobson, M.Z. 2002. Atmospheric Pollution: History, Science, and Regulation. Cambridge University Press, New York.

U.S. Environmental Protection Agency (US EPA). 1998. An Office Building Occupant's Guide to Indoor Air Quality (EPA-402-K-97-003). GPO, Washington, DC.

NINETEEN

Soil Characterization

OBJECTIVES

- Be able to identify soil horizons in the stratigraphy of a soil test pit
- Be able to use a color and texture guide to describe soil characteristics
- Be able to describe the process of digging a soil test pit
- Be able to identify the components of a soils report

KEY CONCEPTS & TERMS

- ✓ Pedosphere
- ✓ Soil horizon
- ✓ Parent material
- ✓ Soil disturbance
- ✓ Soil profile
- ✓ Soil texture triangle

INTRODUCTION

Soil is a thin layer, called the **pedosphere**, on top of most of Earth's land surfaces. This thin layer is a precious natural resource. Soils so deeply affect every other part of the ecosystem that they are often called the "great integrator." Soils hold nutrients and water for plants and animals. Water is filtered and cleansed as it flows through soils. Soils affect the chemistry of the water and the amount of water that returns to the atmosphere to form rain. The foods we eat and most of the materials we use for paper, buildings, and clothing are dependant on soils. Understanding soil is important in knowing where to build our houses, roads, and buildings as well as understanding environmental impacts.

Soils are composed of three main ingredients: minerals; organic materials from the remains of dead plants and animals; and pores that may be filled with air or water. A good quality soil for growing plants should have about 45% minerals, 5% organic matter, 25% air, and 25% water. Soils are dynamic and change over time. Some properties, such as soil moisture content, change very quickly (over hours), while other changes, such as mineral transformations, occur very slowly (over thousands of years).

Soil formation (pedogenesis) and the properties of the soils are the result of five key factors. These factors are:

1. **Parent material:** the material from which the soil is formed. Soil parent material can be bedrock, organic material, or surficial deposits from water, wind, glaciers, or volcanoes.

2. **Climate:** heat and moisture break down the parent material and affect how fast or slow the soil processes go.

3. **Organisms:** all plants and animals living on or in the soil. The dead remains of plants and animals become organic matter in the soil, and the animals living in the soil affect the decomposition of organic materials.

4. **Topography:** the location of a soil on the landscape can affect how climatic forces impact it. For example, soils at the bottom of a hill will be wetter than those near the top of the slopes.

5. **Time:** all the above soil-forming factors assert themselves over time, from hundreds to tens of thousands of years.

SOIL PROFILES

Due to the interaction of the five soil-forming factors, soils differ greatly. Each soil on the landscape has its own unique characteristics. The way a soil looks if you dig a hole in the ground is called a **soil profile**. The soil profile can be used to determine the properties of the soil and the best use of the soil. Every soil profile is made up of layers called **soil horizons**. Horizons can be identified by changes in color or texture compared to adjacent horizons. Horizons are labeled based on their properties (Figure 19.1).

O Horizon: the O-horizon is made up of organic material. The horizon is found at the soil surface.

A Horizon: the A-horizon is commonly called the topsoil and is the first mineral horizon in the soil profile. The A-horizon is mostly made up of sand, silt, and clay particles, but also contains some decomposed organic material.

B Horizon: the B-horizon is composed of mineral material which is undergoing chemical and physical weathering. Weathering causes changes in soil color, texture, and structure. The B-horizon is often rich in clays, iron, and aluminum.

C Horizon: the C-horizon is the parent material from which the horizons above have formed.

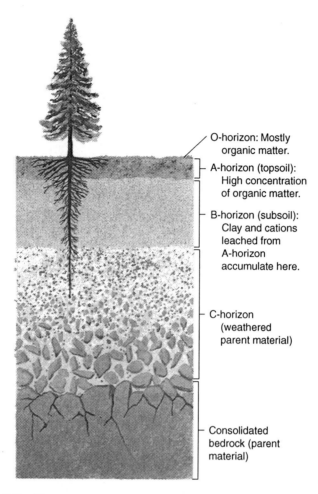

O-horizon: Mostly
organic matter.

A-horizon (topsoil):
High concentration
of organic matter.

B-horizon (subsoil):
Clay and cations
leached from
A-horizon
accumulate here.

C-horizon
(weathered
parent material)

Consolidated
bedrock (parent
material)

Figure 19.1—Typical Soil Profile.
This profile in an area of high soil deposition is not to scale, but assume the tree is just a few feet tall, and note that the root structure may be quite different from the above-ground plant. Horizon depths can vary greatly depending on flooding and other factors. Source: Raven, P.H. and L.R. Berg. 2004. Environment. 4th ed. John Wiley & Sons, NY (p. 312).

SOIL CHARACTERIZATION

In the field, soil pits are routinely dug, and soil horizons are characterized according to color, texture, structure, consistence, amount of roots and rocks. In this lab we will be identifying soil horizons and determining soil color, soil texture, and percent of roots and rocks.

Color: the color of the soil changes depending on how much organic matter is present and the kinds of minerals it contains. Soil color will differ depending on moisture content, and the color can often indicate if the soil has been saturated with water.

Texture: the texture is the amount of sand, silt, and clay particles in the soil and can be determined by how the soil feels. Sand is the largest size particle in the soil and feels gritty. Silt feels smooth or floury. Clay feels sticky.

MATERIALS

- ☐ Meter stick or tape
- ☐ String and string level
- ☐ Munsell color book
- ☐ Shovel
- ☐ Trowel
- ☐ Sturdy shoes and outdoor clothes
- ☐ Water bottle

TASK

1. Dig a pit one-meter deep and roughly 1-meter square. A pit this big is needed to easily observe all the soil horizons. Note that you will need to reclaim the site, so be neat and careful with the removed soil.

2. Starting at the top of the soil profile, observe the profile closely to determine where the different horizons occur. Look carefully for any distinguishing characteristics such as different colors, roots, and the amounts and size of rocks, and so on.

3. Mark the horizon boundaries with nails, and measure the top and bottom depths for each horizon, in cm, and record on a piece of graph paper (depict the horizons, measurements, and observations).

4. Assign a horizon label (O,A,B,C) to each horizon.

5. Characterize <u>each</u> horizon for (a) color using a Munsell color book, (b) texture using the soil texture flowchart and the soil textural triangle, and (c) percent and size of roots and rocks.

   a. **For color:** take a sample from the horizon and moisten it slightly with water from your water bottle. Hold the color chart next to the soil, and determine which color matches the color of your soil. Stand with the sun over your right shoulder so the sun shines on the color chart and your soil sample. Record the color on your soil characterization sheet. Sometimes a soil horizon may have more than one color. Record all colors.

   b. **For texture:** See Figures 19.2 and 19.3.

   c. **For roots and rocks:** Estimate the percent and approximate size of roots and rocks.

6. Other site information—Spend a few minutes recording the details of the site.

   a. Record the dominant vegetation at the site.

   b. Record the percent slope of the land (rise ÷ run).

   c. Record the location of the soil pit using major features, such as buildings, utility poles, etc.

   d. Record any other distinguishing properties of the site.

   e. Record time of day, date, approximate temperature, and weather.

   f. Document evidence of previous disturbance (fill, burrowing) in your pit and on the site.

   g. What biotic activity can you see in the walls of your test pit?

7. Completely restore the site so that in a few weeks no one could tell a pit was dug.

LAB WRITE-UP

The lab write-up is not a formal lab report (in the sense of an experiment), but should be written as a professional report and <u>must</u> include (a) a clean version of your soil characterization on graph paper and (b) answers to the following questions:

1. What are the color, texture, and percent and size of roots and rocks for each soil profile?

2. What types of plants and animals do you find in your soil pit and in the general areas around your site? (You will need to do some research.)

3. How do they influence soil formation?

4. What is the parent material from which the surveyed soil formed?

5. What is the general climate at your soil site?

6. What is the recent land use in this area?

7. What other potential land uses do you think your soil could support? Why?

8. What do you think has been the dominant soil-forming factor(s) at your site?

9. How do the abiotic factors of soil influence the types of trees and plants that live in a particular area?

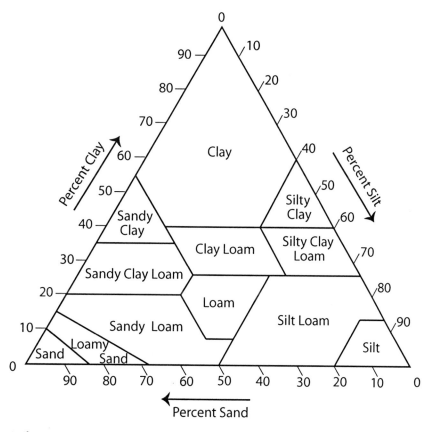

Figure 19.2—Soil Texture Triangle.
This triangle illustrates the twelve USDA soil classifications and is commonly used in classifying soil textures.
Source: U.S. Department of Agriculture. 1993. Soil Survey Manual. Handbook No. 18. Washington, DC.

Step 1

a. Obtain and moisten a sample from each Horizon.
b. Place some soil from a horizon (about the size of a small egg) in your hand.
  i. Using a spray bottle, moisten the soil.
  ii. Let the water soak in and then work the soil between your fingers until it is moist throughout.
  iii. Try to form a ball.
  iv. If the soil forms a ball, go to **Step 2**.
  v. If the soil does <u>not</u> form a ball, go to **Step 5**.

Step 2

A. If the soil: • Is really sticky • Is hard to squeeze • Stains your hands • Has a shine when rubbed • Forms a long ribbon (5+ cm) without breaking Classify it as *clay* and go to **Step 3**; otherwise, go to **B**.	**C.** If the soil: • Is soft • Is smooth • Is easy to squeeze • Is at most slightly sticky • Forms a short ribbon (< 2 cm) Classify it as *loam* and go to **Step 3**; otherwise, go to **D**.
B. If the soil: • Is somewhat sticky • Is somewhat hard to squeeze • Forms a medium ribbon (between 2 and 5 cm) Classify it a *clay loam* and go to Step 3; otherwise, go to **C**.	**D.** If the soil forms a ball but no ribbon, go to **Step 4**.

Step 3

Refine the initial soil texture classification from **Step 2** for relative amounts of *sand* and *silt*. Wet a small pinch of soil in your palm and rub it with a forefinger. If the soil: feels gritty, go to **E**; very smooth with no gritty feeling, go to **F**; or only a little gritty, go to **G**.	**E.** Add the word *sandy* to the initial classification Soil texture is (select one) • Sandy clay • Sandy clay loam • Sandy loam Soil Texture is complete.
F. Add the word *silt* or *silty* to initial classification. Soil texture is (select one) • Silty clay • Silty clay loam • Silt loam Soil Texture is complete.	**G.** Leave the original classification of (select one) • Clay • Clay loam • Loam Soil Texture is complete.

Step 4

(Test for *loamy sand* or *silt*)

If the soil:
- Forms a ball
- Forms no ribbon
- And is:

H. Very gritty (otherwise go to **I**)
Soil texture is *loamy sand*
Soil texture is complete.

I. Very soft and smooth with no gritty feeling.
Soil texture is *silt*.
Soil Texture is complete

Step 5

Test for Sand

If the soil does not form a ball and instead falls apart in your hand,

Soil texture is *sand*.
Soil Texture is complete.

Figure 19.3—Steps for Classifying Soil Texture.
Working your way through these steps with a soil sample should lead to successful classification of texture. Adapted from the U.S. Department of Agriculture. 1993. Soil Survey Manual. Handbook No. 18. Washington, DC. and U.S. GLOBE Program. Texture by Feel Guide. [Online] Available at http://ltpwww.gsfc.nasa.gov/globe/tbf/tbfguide.htm (verified 10 June 2004).

TWENTY

Environmental Assessment and Project Planning: Desktop Research

OBJECTIVES

- Be able to read and interpret standard geographical, historical, and natural resource maps
- Be able to use a variety of maps to determine what areas are best suited for land use development

INTRODUCTION

Many people who do planning or environmental impact assessment begin their task by exploring maps. Desktop references are a valuable source of information to catalog natural resources and to assess potential impacts from human activities. This lab will familiarize you with these maps and provide an opportunity to discover what types of information can be gleaned from them by assessing the environmental viability of a project by using desktop references. Much of this work is done in the environmental consulting world with Geographic Information Systems (GIS), but the principle is the same.

MATERIALS

- ☐ Soil survey maps
- ☐ Surficial geology maps
- ☐ National wetland inventory maps
- ☐ Topographic maps
- ☐ Aerial photographs
- ☐ Sanborn Fire Insurance maps
- ☐ Other maps as available (e.g., road map, bedrock geology)
- ☐ Ruler, graph paper

SOIL SURVEY MAPS

The U.S. Department of Agriculture's National Cooperative Soil Survey (NCSS) is a county-by-county scientific inventory of U.S. soils on nearly all public and private land. A soil survey includes soil maps and descriptions of each type of soil in the county. Maps show the location of soils in a county. Descriptions of each soil type include:

- Depth of each major soil layer.
- The ability of water to infiltrate the soil and how easily roots can penetrate.
- The rate at which water moves downward through the soil.
- How much water the soil can store for plants.
- How acid or alkaline the soil is.
- The soil's susceptibility to erosion by water and wind.

The soil survey is used to identify the suitability of soils for the construction of buildings, roads, septic tank absorption fields, sewage lagoons, landfills, ponds, and dikes and levees.

SURFICIAL GEOLOGY MAPS

Surficial geology maps depict surficial rocks and sediments. Because in most environments, vegetation, soils, and human structures cover the surface, the underlying rocks and sediments are not directly visible or exposed. Surficial geology maps depict the materials approximately 5 feet below the surface.

NATIONAL WETLAND INVENTORY MAPS

The National Wetlands Inventory (NWI) of the U.S. Fish & Wildlife Service produces information on the characteristics, extent, and status of the nation's wetlands and deepwater habitats. Approximately 90 percent of the lower 48 states and 34 percent of Alaska have been mapped. NWI maps are used in a myriad of applications including planning for watershed and drinking water supply; siting of transportation corridors; construction of solid waste facilities; siting of buildings, wildlife habitat identification; floodplain planning; and endangered species recovery.

TOPOGRAPHIC MAPS

The U.S. Geological Survey produces topographic maps of various scales (1:24,000 is common). Note that since the scale is a ratio, it does not matter if it is in metric or English units, although information is generally given in both. The feature that most distinguishes topographic maps from other maps is the use of contour lines to portray the shape and elevation of the land. Topographic maps render the three-dimensional changes of terrain onto a two-dimensional surface. (However, as shown in Figure 20.1, the slope can be determined from topographic maps.) Topographic maps show and name **natural features** including mountains, valleys, floodplains, lakes, rivers, wetlands, and forests. They also identify human-made artifacts, such as roads, boundaries, transmission lines, and major buildings. Topographic maps are used for land use planning, energy exploration, natural resource conservation, environmental management, public works design, and outdoor activities. In addition, historical topographic maps can be used to assess growth and temporal impacts to natural resources.

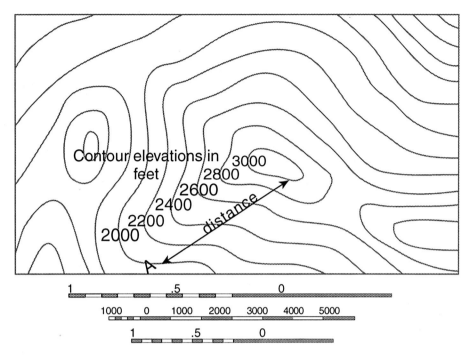

Slope $= \dfrac{\text{Rise}}{\text{Run}}$

% slope $= \dfrac{\text{Rise}}{\text{Run}} \times 100$ **Slope angle** $= \arctan\left(\dfrac{\text{Rise}}{\text{Run}}\right)$

Run = horizontal distance from A to B = 1 mile = 5280

Rise = elevation change = 3000 - 2000 = 1000 ft.

% slope $= \dfrac{1000 \text{ ft}}{5280 \text{ ft}} \times 100 = 18.9\,\%$

Slope angle $= \arctan\left(\dfrac{1000 \text{ ft}}{5280 \text{ ft}}\right) = 10.7° \text{ slope}$

Figure 20.1—Calculating Slope Using a Topographic Map.
In this example English units are used primarily because elevations on USGS maps are provided in English units. However, for scientific use, the metric system is preferred. The formulae are the same for either case.

AERIAL PHOTOGRAPHS

Aerial photographs are used to prepare topographical maps and are especially important in depicting land use and land cover. The U.S. Geological Survey began using aerial photographs in the 1930s to construct topographical maps. Historical aerial photographs are also excellent sources to trace land use changes and identify previous industrial operations, which can help identify potential areas of contamination. Most of the photos are black and white, but some are color-infrared that are better to show certain differences in plant life. Aerial photos are shot from airplanes flying at a constant altitude of 20,000 feet, and each is shot straight down from the plane. Each 9 × 9 inch print covers an area about 5 miles square at an approximate scale of 1:40,000, where an inch represents about 0.6 mile on the ground.

SANBORN FIRE INSURANCE MAPS

Sanborn Fire Insurance maps were published from the mid- to late 1800s to about 1980. They are a valuable resource to evaluate the potential for past contamination based on historical uses of urban property. Historically, some of the most serious fire risks were related to chemical, oil, and gas storage. Consequently, their location, size, and contents are noted prominently on historic fire insurance maps. In addition, insurance maps also note the name and/or type of manufacturing operations, which may indicate a potential contamination problem as shown in Table 20.1. Until 1980, waste management practices were generally substandard. Waste was placed on the ground, discharged through substandard sewers, or stored in temporary devices. Consequently, waste contaminants may have leached and contaminated the soil and/or groundwater.

TABLE 20.1   **Manufacturing Activity and Likely Associated Chemical Pollutants**

Manufacturing Activity	Likely Chemical Pollutants
Leather tanning	Heavy metals, cyanide, chromium
Metal foundry	Heavy metals, solvents oil, acids
Paint	Solvents, lead
Silverware	Heavy metals, cyanide, solvents, acids
Kerosene oil works	Oil, kerosene
Gas works	Oil, PNAHs (polynuclear aromatic hydrocarbons)
Photography	Mercury, silver, solvents, acids

TASK

■ As a team, you will be assigned one of the following structures:

___ A 100-acre landfill containing lined cells for long-term, stable waste storage of non-hazardous solid waste. This must be no closer than 1 km to any surface waters.
___ A cluster of 20 single-family homes on 100 acres. Assume it is desirable to be within 10 km of a pond, lake, or river.
___ A temporary (but stabilized and monitored) low-level radioactive waste storage site on 150 acres.
___ A 25-acre chemical manufacturing plant
___ A 30-acre shopping mall
___ Other: _____

■ On graph paper, draw and then cut out the project's footprint (assume the footprint of the project will be a square or rectangle) using the appropriate scale for the USGS Quad map.

■ Review all the provided maps and references. Using the different maps and your project tract outline, determine an optimal location for your assigned project. (See Figure 20.2 for a conceptual application of the maps to your site project.)

■ For each site, you will need to consider its location (i.e., distance, direction, and slope) to the nearest: population, surface water, wetlands, and natural resource features; suitability of the soil; slope; and depth to groundwater.

■ The acceptability of the physical location depends on the operations conducted at the site. For example, the landfill and radioactive waste facility need to be away from surface and ground water, the chemical manufacturing plant needs to be away from populations and protected areas but near a supply source of water, and the housing development and shopping mall need to be near roads to reduce secondary environmental impact.

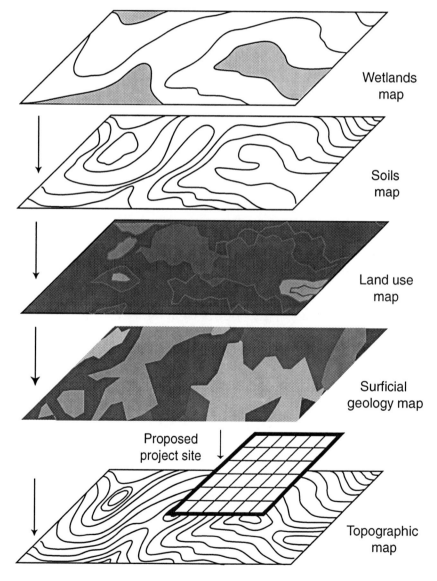

Figure 20.2—Conceptual Application of the Maps to the Site Project.
This type of overlay can be done with actual maps, with Geographic Information Systems (GIS), or mentally.

WRITE-UP

The write-up for this lab is a project evaluation. Basically, you will be writing a two- to three-page evaluation of the proposed project using the following headings. (Show all work and calculations in footnotes, not just your answers.)

Project Headings and content:

1. Title of Project

2. Project Type

3. Reference Location

   a. Provide the grid reference points for each of the four corners of your project on the 7.5' USGS Topographical maps. You should be able to tell from the scale that 70 miles is equal to 1 degree; 1 minute (60 minutes in a degree) is equal to 1.2 miles; and 1 second is equal to .02 miles. Locate the latitudinal and longitudinal reference lines on the map. Measure from the reference point to the project boundaries remembering to add or subtract as necessary.

   b. Be sure to reference the proper USGS quad map name.

   c. What is the scale of the Quad map in English and metric (1 cm on the map = how many meters on the ground)?

4. Site Characteristics should include:

   a. Land cover and current use in the area and on the site

   b. Dominant soil types on the site

   c. Type, size, and distance of nearest wetland

   d. Elevation (metric and English) of highest and lowest point (and the names) of your selected site

   e. Average slope of the project area (or slope characteristics if the area is many hectares)

   f. Distance, direction, and slope (e.g., down- or up-slope) to nearby natural resource features and descriptions of these features

   g. Distance and direction to the nearest settlement

   h. Type, size, distance, direction, and slope to closest water course (i.e., river, brook, or stream)

   i. Type, size, distance, direction, and slope to closest non-wetland surface water (i.e., lake, pond, or ocean)

5. Site Suitability

   a. Describe the suitability of the site based on your answers to question 4. Be sure to support your choice for where to place the site.

   b. What other factors not represented or depicted on the maps should be considered regarding the selection of your project?

6. Lab Evaluation

    a. Critique the information provided by the maps in the context of having to make a decision about where to put your project.

    b. Were there a lot of choices or very few?

    c. Was it difficult to find an ideal location?

    d. What other information would you have wanted to have, and what other kinds of maps would be good for the assigned task?

    e. What did you learn by doing this?

Pollution Prevention: Solid Waste

OBJECTIVES

- Be able to explain and apply the Pollution Prevention Hierarchy
- Be able to collect solid waste data
- Be able to survey local attitudes about solid waste

KEY CONCEPTS & TERMS

- ✓ Municipal solid waste
- ✓ Pollution prevention
- ✓ Pollution Prevention Act of 1990
- ✓ Reduce
- ✓ Reuse
- ✓ Recycle

INTRODUCTION

Americans are the most wasteful society on the planet. In 1960, the U.S. per capita waste generation rate was 2.7 pounds per person per day. Currently, we generate an average of 4.3 pounds per person of **municipal solid waste** (i.e., trash, garbage, rubbish, and refuse) every day (US EPA, 2003). Municipal solid waste includes paper, yard waste, plastics, metals, wood, glass, and a variety of other materials—a typical household generates enough each year to fill an entire house.

What action can be taken to reduce the amount of waste? An approach to reduce the amount of waste generated is to adopt and apply the Pollution Prevention Hierarchy (Figure 21.1).

THE POLLUTION PREVENTION HIERARCHY

The Pollution Prevention Hierarchy was established by Congress in Section 6602(b) of the **Pollution Prevention Act of 1990**. In this Act, Congress created a national policy that:

- Pollution should be prevented or reduced at the source whenever feasible;
- Pollution that cannot be prevented should be reused/recycled in an environmentally safe manner whenever feasible;
- Pollution that cannot be prevented or recycled should be treated in an environmentally safe manner whenever feasible; and

● Disposal or other release into the environment should be employed only as a last resort and should be conducted in an environmentally safe manner.

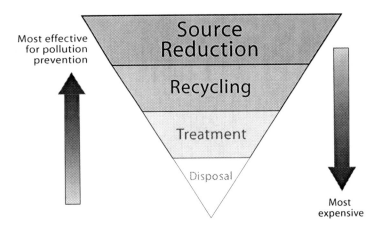

Figure 21.1—The Pollution Prevention Hierarchy.
This hierarchy depicts the preferential approach to environmental management by preventing the creation of pollution rather than managing pollution after it is generated. Thus, source reduction is the most preferred method to prevent pollution, and disposal is the least preferred method. Correspondingly, source reduction is the least expensive approach compared to disposal. Note that the cost is not limited to direct costs, but includes cleanup costs, social costs (impacts on health), and environmental damage costs.

SOLID WASTE MANAGEMENT

As a society, we seem to have difficulty in the simple act of recycling (let alone reducing and reusing waste). It is common practice to combine all our solid waste and let someone else worry about sorting it for processing or reclamation. Generally this consolidation practice creates an economic and technological barrier to increased reclamation. It generally takes more money, energy, and resources to separate trash after it is generated rather than to properly sort it before it is generated. Why do we manage waste in this ineffective and inefficient way? How closely does the pyramid at the left in Figure 21.1 resemble our present means of dealing with waste?

Much of our solid waste is "disposed" of in landfills or is incinerated. We landfill or burn many things that could be recycled. In recent years, there has been a sustained effort to divert waste away from disposal and recapture resources or energy through recycling. As shown in Figure 21.2, 30% of the nation's municipal solid waste is recycled, which represents a doubling since 1990. Source reduction prevents the generation of waste or pollution. Reuse is when a waste product or pollutant is used again without substantial processing. Most recycling involves the substantial alteration or reprocessing of a product or pollutant. Recycling is less preferable than reduction or reuse, but much better than simply dumping, hiding, or burning our waste.

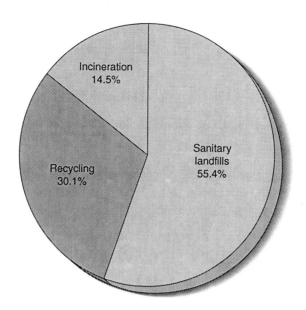

Figure 21.2—Disposal of Municipal Solid Waste in 2000.
The percentage of recycled materials has been gradually increasing. Source: Raven, P.H. and L.R. Berg. 2004. Environment. 4th ed. John Wiley & Sons, NY (p. 246). Data from the U.S. Environmental Protection Agency (http://www.epa.gov/epaoswer/non-hw/muncpl/facts.htm).

MATERIALS

No specialized materials required

TASKS

In teams, conduct a thorough walk-through of your building by walking down each hall on each floor to answer the following questions:

1. In a table, list <u>all</u> waste types generated in your building. In a separate column of the table, list the source of generation for the waste. (See Table 21.1 for an example.)

TABLE 21.1	University Sources of Waste
Waste	**Sources**
Paper	Classrooms . .
Food	Vending machine, snack bar . . .

2. Count the number and type of trash <u>and</u> recycling containers in the hallways of your building and input into a table. (See Table 21.2 for an example.)

TABLE 21.2	Results of the Solid Waste Survey		
Location	**Collection Container Type**	**Contaminated**	**Inappropriate Materials**
1st Floor, near room 1xx	Paper	Yes	food, plastic
	Trash	Yes	paper

3. Visually inspect the containers (don't root through containers and risk personal injury).

   a. What is the frequency of "contamination" in percent?

   b. What types of non-recyclable materials are in the collection container?

   c. What types of recyclable materials are in the trash container?

4. During your inspection of the building:

   a. Count the number and type of posters, signs, and education/communication materials that explain why we should recycle, where we should recycle, and how (e.g., what types of materials go where).

   b. Note their location. Are they close to the waste generation point or recycling container collection point?

5. Go to a food service operation (e.g., snack bar, cafeteria) and conduct a visual inspection.

   a. Are they attempting to promote source reduction, reuse, or recycling—include waste food—if yes, how; if no, how)?

   b. What percentage, roughly, of the non-food items is reusable?

   c. What percentage would you estimate is recyclable?

6. Select a <u>random</u> sample (at least 10) of students, staff, and faculty. Survey them by asking the following questions:

   - Do you recycle?
   - What percent of all waste generated at the university is recyclable?
   - What waste materials generated at the university do you believe are recyclable?
   - Do you believe there are an appropriate and sufficient number of containers for recycling?
   - We have found numerous instances of trash in recyclable containers and recyclables in trash containers. What two suggestions can you make to reduce this practice?

   a. Tabulate <u>and</u> graph your results.

   b. What conclusions can you draw from your sample?

   c. How did you choose your sample?

7. What does a recycling container cost to purchase? What does it cost to monitor the container and empty it? In your opinion, how has this affected recycling?

8. In the conclusion to your write-up, answer the following questions:

   a. Is your university doing the best possible job in promoting pollution prevention? (If yes, what evidence do you have to support this answer? If no, what could be done to improve compliance?)

   b. What could be done to reduce the degree of non-compliance with "contaminated" receptacles?

   c. What overall conclusions can you draw regarding students and environmental awareness (three-sentence minimum)? What could be done to improve awareness?

9. List at least four Internet sites that could serve as resources for promoting a "greener" campus.

Reference

U.S. Environmental Protection Agency (US EPA). 2003. Municipal Solid Waste in the United States, 2001 Facts and Figures, EPA 530-S-03-011, Washington, DC.

TWENTY-TWO

Field Trip: Solid Waste Management

OBJECTIVES

- Be able to describe how local municipal solid waste is managed
- Be able to estimate the volume of trash generated by local communities

INTRODUCTION

Historically, solid waste management has generally been a local responsibility. Local governments have relied on landfilling as the cheapest disposal method, depicted in Figure 22.1, but it also has the greatest environmental impact. Incineration can represent a better alternative, but it is costly. Usually, a preferred method is waste-to-energy (WTE), which uses combustion of the waste to produce electricity as a byproduct as shown in Figure 22.2. However, it too is expensive. Both incineration and WTE reduce the volume of trash by 90% but the resulting ash must be disposed of properly. Of course the best approach is to reduce, reuse, and recycle waste to avoid its disposal, but Americans have embraced this practice—pollution prevention—to only a small degree. The more we know about solid waste management, the better we can understand solid waste and reduce its impacts.

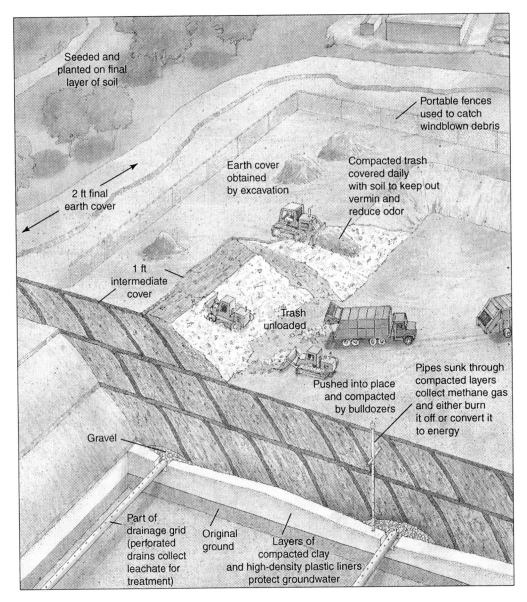

Labels in figure:
- Seeded and planted on final layer of soil
- Portable fences used to catch windblown debris
- Earth cover obtained by excavation
- Compacted trash covered daily with soil to keep out vermin and reduce odor
- 2 ft final earth cover
- 1 ft intermediate cover
- Trash unloaded
- Pushed into place and compacted by bulldozers
- Pipes sunk through compacted layers collect methane gas and either burn it off or convert it to energy
- Gravel
- Part of drainage grid (perforated drains collect leachate for treatment)
- Original ground
- Layers of compacted clay and high-density plastic liners protect groundwater

Figure 22.1—Municipal Solid Waste (MSW) Landfill.

Modern landfills are underlain with a monitored protective liner of high-density plastic or clay. Generally, above this layer a leachate collection and removal system is installed to remove liquid leachate. Following each day of operation, the landfill is covered. Soil used to be the preferred cover, but recyclable materials and lightweight foams are now used. Source: Raven, P.H. and L.R. Berg. 2004. Environment. 4th ed. John Wiley & Sons, NY (p. 547), and U.S. Environmental Protection Agency (EPA). Landfill Methane Outreach Program [Online]. Available at http://www.epa.gov/lmop.

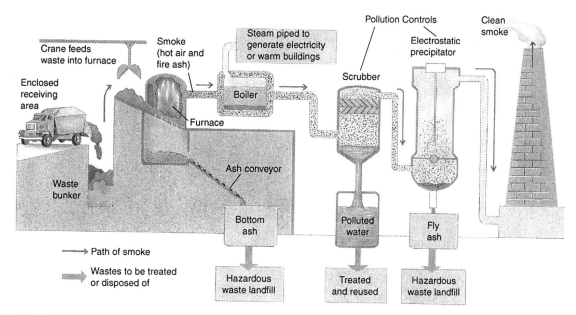

Figure 22.2—Mass Burn, Waste-to-Energy Incinerator.
Scrubbers and electrostatic precipitators help trap noxious emissions. Fourth in the world, the U.S. burns 16% of its waste in waste-to-energy plants (1% is burned in old-fashioned incinerators that do not create energy). The number one waste-to-energy country is Japan (62%). Source: Raven, P.H. and L.R. Berg. 2004. Environment. 4th ed. John Wiley & Sons, NY (p. 550) and Energy Information Administration Annual Energy Review. EIA, Washington DC. (see http://www.eia.doe.gov/ for current energy information).

QUESTIONS TO ADDRESS IN RESPONSE TO THIS FIELD TRIP

1. How much waste does the facility handle?

2. What geographic area does it serve?

3. What type of waste management operations does the plant employ? What are the environmental impacts?

4. Are there seasonal differences in the amount of waste generated?

5. How does waste generation relate to sustainable development?

6. How can we reduce waste generation? Why is recycling not a complete answer to our waste problems?

7. Why is the recyclable materials market so volatile?

8. Is there a sector of our economic system that seems particularly important in determining the levels of solid waste we generate? Are there any solutions to this issue?

9. Create a visual product of your experience, something that could be used by the facility's staff. For example, make a diagram, chart, or graph to show the amount and/or type of waste processed by this facility.

CHAPTER TWENTY-THREE

Toxicity Testing and the LC$_{50}$

OBJECTIVES

- Be able to count and categorize organisms using a dissecting microscope
- Be able to describe a method used in testing for product toxicity
- Be able to create a concentration-response curve for toxicity

TERMS & KEY CONCEPTS

✓ Concentration Response Curve
✓ Lethal Concentration 50%
✓ Lethal Dose
✓ Toxicity

INTRODUCTION

How do government and industry determine acceptable levels of human exposure to cosmetics, household chemicals, pesticides, and new chemical products? A major tool is through toxicity testing to generate a concentration-response curve. In this lab, you will become familiar with the methodology of generating a concentration-response curve (CR curve) and investigate the advantages and disadvantages of this approach. A CR curve (a variation of the dose response curve depicted in Figure 23.1) is used to assess the effects of various concentrations of a chemical substance on a group of organisms. Notice that the term "chemical substances" is used instead of "toxic substances." Every chemical is toxic; it is the concentration that makes the poison. For example, table salt will cause an adverse effect in humans, but it takes a relatively large concentration to do this. In contrast, only very small amounts of cyanide are necessary.

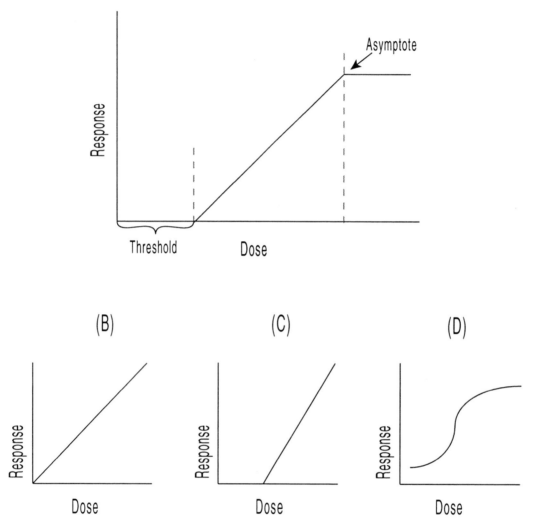

(A) General Response Curve

Figure 23.1—Generalized Response Curve.

(A) An example of a general response curve. The *threshold level* is the level at which an adverse effect is not observed. The *asymptote* represents the maximum effect (response) level. (B) An example of a *linear response curve* where a unit increase in the dose produces a unit increase in the response. This curve has no threshold level. Current national policy assumes there is no "safe" dose of carcinogens, thus, there is no threshold. (C) An example of a *linear response curve* with a threshold level. Thus, there is a certain dose where no adverse effect is observed. (D) An example of a *non-linear response curve*, in which the shape of the relationship is curved. Thus, a unit increase in dose will have varying responses. Some nonlinear response curves have a threshold and/or an asymptote—when the dose reaches an asymptote, no additional dose affects the response.

To estimate toxicity, a substance is tested at various concentrations with living test organisms to determine what concentrations elicit a response in the organism. Numerous tests exist, many of which are conducted on animals. Selection of a test depends on many factors, including the potential target organ, whether it is a **chronic** (long term, such as cancer, liver disease) or **acute** (short term, such as poisons) response or condition test, and available resources (Figure 23.2). Chemical toxicity testing can be very expensive.

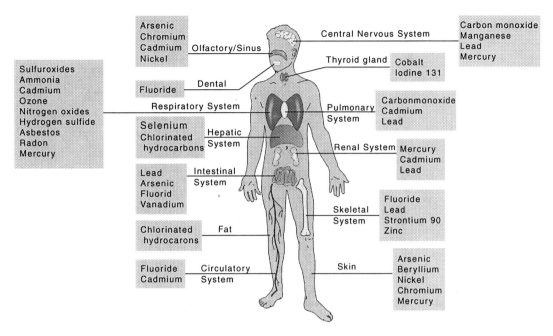

Figure 23.2—Target Organs and Pollutants.
The Effects of Some Major Pollutants in Humans. Source: Bodkin, D.B. and E. A. Keller. 2003. Environmental Science: Earth as a Living Planet. 4th ed. John Wiley & Sons, NY. (p. 298) and Waldbott, G.L. 1978. Health Effects of Environmental Pollutants. 2nd ed. Moseby, St. Louis.

One standard measurement of the acute toxicity of a substance is the lethal concentration 50% (LC$_{50}$): the concentration that causes death to 50% of the test organisms. The lower the LC$_{50}$ value for a substance, the greater the toxicity.

Measuring toxicity requires a toxicologist to plot data in the form of a concentration-response curve. This curve relates the concentration of the chemical to the percentage of animals showing the response (e.g., death). This curve will allow you to determine the concentration of a toxic material that causes 50% mortality in a population of test animals. Although the goal is to establish the 50% level, knowing the shape of the curve above and below the midpoint is important. A hypothetical example of a concentration-response curve is depicted in Figure 23.3, but note that the response might be much more linear or it might have a lower threshold.

The response to the concentration is simply the amount of damage it causes. There may be visual or measurable symptoms of toxicity at sub-lethal levels. The most common endpoint for most chemicals tested is death because it is easily identifiable, reliable, and cheaper than sub-lethal assessments.

Concentration-Response Curve of a Chemical

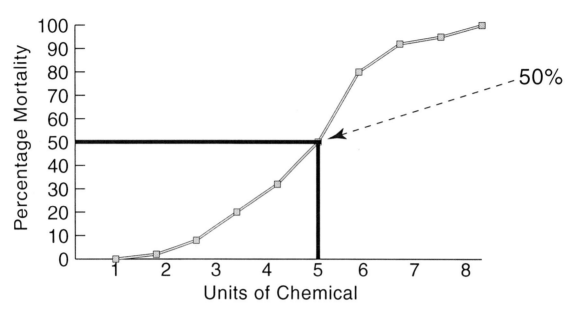

Figure 23.3—A Hypothetical Concentration-Response Curve LC_{50}.
The curve may be quite different depending on the substance and the measurements. Compare with the dose-response curve (Figure 23.1).

Generally, researchers do not intentionally use human subjects for testing the toxic effects of chemicals. Common test organisms include zooplankton, algae, mice, rats, and dogs (rats and mice account for 90% of the animals used in toxicity testing); the results are then extrapolated to humans to estimate a "safe" exposure level. Because of our vastly increased knowledge of toxicity, and the ethical concerns with testing on animals, federal agencies have reduced the reliance on animal testing.

In this lab we will perform a simple experiment to examine the effects of a substance on brine shrimp (*Artemia franciscana*) sufficient to calculate a concentration-response curve.

MATERIALS

- ☐ 5 petri dishes per group (either square dishes with etched grids or round dishes underlain by graph paper)
- ☐ Brine shrimp
- ☐ Brine (NaCl and water) solution
- ☐ Vinegar or other contaminant (e.g., ammonia, oil, household pesticide)
- ☐ Pipettes
- ☐ Dissecting microscopes
- ☐ Labels
- ☐ Beakers
- ☐ Graduated cylinders

TASKS

In teams, conduct the following tasks. As individuals, prepare a formal report on the results:

Step 1

Write a hypothesis. Remember, you are hypothesizing what the LC$_{50}$ is (x%).

Step 2

Obtain five petri dishes. Place 16 ml of prepared NaCl solution into each petri dish.

Step 3

Label each petri dish by vinegar concentration (see table below). Do not put the vinegar in the dish yet.

Step 4

Place four (4) ml of brine shrimp solution into each petri dish.

Step 5

Count the total number of brine shrimp in each petri dish. Count the numbers of dead and alive (do not count the unhatched eggs) so that you will be able to tell how many dead are attributable to the vinegar.

(You can estimate the number by counting the number of shrimp in at least 16 randomly[15] selected squares, calculate the per-square average, and multiply the average by the number of whole squares. However, this method is valid <u>only</u> if the brine shrimp are <u>evenly</u> distributed.)

Step 6

Record the time on each label and then add the requisite amount of vinegar (according to Table 23.1 below or choose your own concentrations) in each petri dish. Gently swirl to mix. Wait five minutes, count and record the number of dead.

TABLE 23.1	**Some Contaminant Concentrations for Toxicity Testing with Brine Shrimp**				
Dish	**NaCl Solution**	**Brine**	**Contaminant**	**Calculate**	**Concentration**
*1	16 ml	4 ml	0 ml	0/20	0.0%
2	16 ml	4 ml	.5 ml	.5/20.5	2.4%
3	16 ml	4 ml	1 ml	1/21	4.75%
4	16 ml	4 ml	2 ml	2/22	9.0%
5	16 ml	4 ml	3 ml	3/23	13.0%

*This is your control.

[15]You can use a random number table (commonly found in the back of mathematics and statistics books) or search online for a random number generator such as the one at http://www.graphpad.com/quickcalcs/randomN1.cfm.

Step 7

Determine the percent brine shrimp that died in each petri dish. (Number of dead in Step 6 minus the number of dead in Step 5.)

Step 8

Do your results allow you to accept or reject your hypothesis? That is, did you find the LC$_{50}$? Remember, you are trying to accomplish two things in this lab: identification of the LC$_{50}$ and to construct a dose-response curve. Based on your results from Step 7, repeat the process, but using different concentrations to help you find the LC$_{50}$ and construct the curve (i.e., use higher concentrations if appropriate to complete the curve between LC$_{50}$ and LC$_{100}$).

Step 9

Based on your results of Step 8, repeat the experiment one more time using different concentrations to produce a more robust dose-response curve.

Step 10

Construct a single concentration-response curve with all data points as shown in Figure 23.4.

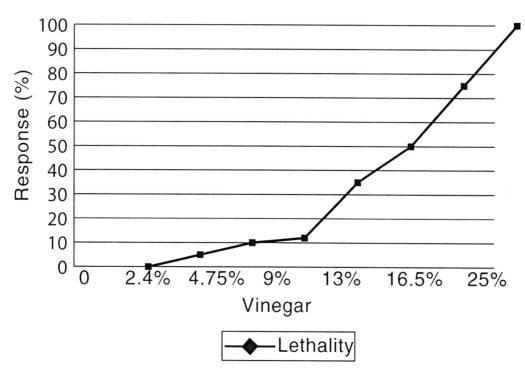

Figure 23.4—The LC$_{50}$ for Vinegar.

In this hypothetical case, the LC$_{50}$ is 16.5% vinegar.

Written Assignment

Each student must submit a formal lab report, which is to include a CR curve in the *results* section. In your *discussion* section, also be sure to discuss the following issues:

1. Problems with your experimental design.

2. Difficulties in extrapolating your lab data to real-world situations (i.e., using your data to determine a safe level for human exposure).

3. Based on your results, try to guess the safe level for human exposure to vinegar (i.e., using animal data to extrapolate safe level for humans). Use your imagination. You are being judged on your thought process, not on whether your level is actually safe.

4. Suggest another substance to be tested and describe why.

Unless otherwise instructed, use a computer for your graphs and essay. Also, remember to attach your data table in its raw form.

Instructions for Generating a Concentration-Response Curve in Microsoft Excel

1. In Excel, combine all the data into a single two-column table as shown below:

Concentration (%)	Mortality (%)
0	0
2.4	4
4.75	10.3
9.0	11.4
3.0	18.9
16.5	42.5

2. In Excel, select "Insert" then select "Chart."

3. Under Standard Types, select "Line" then select "Next>."

4. Select the tab marked "Series." Then, under series, select the button "Add."

5. Select the small square to the right of the title, "Values." Because you want the mortality on the Y axis, highlight all the mortality values. Select "Enter," then select "Next>."

6. Select the small square to the right of the title, "Category X-axis labels." Because you want the concentrations on the X axis, highlight all the concentrations, select "Enter" then select "Next>."

7. You can now add the title ("Concentration-Response Curve . . .") and the titles for each axis; then select "Finish."

Environmental Risk Ranking

OBJECTIVES

- Be able to conduct an opinion survey about an environmental issue
- Be able to rank environmental problems in terms of risk and explain the reasoning for the rankings

KEY CONCEPTS & TERMS

- ✓ Environmental risk
- ✓ Exposure
- ✓ Risk perception

INTRODUCTION

Environmental risk is the potential to cause harm to human health and/or the environment. We constantly face environmental risk from the food we eat, the water we drink, the air we breathe, and the soil we touch. For a risk to occur there must be **exposure** (the potential for a person to come into contact with a contaminant). Some risks are more significant than others and depend on a multitude of factors, including toxicity and amount of a contaminant, where we live, how old we are, our daily activities, and contributing factors (e.g., smoking, diet, alcohol).

A major factor in how and what environmental risks we manage depends on how we perceive risks. This factor, **risk perception**, is an individual or group assessment based on feeling or judgment for the potential of an environmental factor or condition to cause harm. For example, most people perceive the risk from hazardous waste to be very high, but most environmental experts rank the risk to be relatively low. In contrast, most people rank the risk of groundwater contamination, a major source of drinking water, from gasoline stored in underground storage tanks to be low, whereas many environmental experts rank this risk much higher. A major reason for this difference is a person's perception, which is based on education, ethnic background, familiarity, past experience, and a variety of socio-economic factors. Risk perception also is heavily influenced by how the media reports on a risk.

MATERIALS

No special materials needed for this lab.

TASK

In this lab, you are going to conduct some original research using a survey to gather data on people's perception of risk.

Below is a list of environmental problems. Note that some appear to be similar but are slightly different. This is one of the challenges of a survey—wording.

1. In your group, select 10 environmental problems from the list. You can list these risks using the suggested template (alternatively, your instructor may supply you with a template or ask you to devise your own). You will need at least 15 copies.

Environmental Problems

Acid deposition	Loss of outdoor recreation
Acid rain	Loss of wilderness
Air pollution	Loss of wildlife habitat
Asbestos	Ozone depletion
*Climate change	Pesticides
Contaminated drinking water	Radon
Drinking water quality	Smog
Food additives	Solid waste
Genetically modified foods	Stormwater runoff
*Global warming	Underground storage tanks
Ground-level ozone	West Nile Virus
Hazardous waste	Urban sprawl
Indoor air pollution	Litter

*These two terms refer to the same environmental problem. As an experiment, try using both terms to see if participants distinguish between the two.

2. Following the creation of the survey, randomly select <u>at least</u> 15 individuals to complete the survey. Unless otherwise instructed, go to a variety of locations and do not limit yourself to your building.

3. Ask each participant to rank the listed environmental problems based on their perception of the risk.

4. It is imperative that you avoid answering specific questions on any environmental problem. By answering questions you will be influencing or educating the participant and thus, skewing your results.

5. After completing the surveys, return to the lab and tally the results. Use percentages and means to interpret your results. In your write-up, answer the following questions:

   a. What was viewed as the highest risk? Are there any discernible patterns (e.g., younger/older, students v. faculty, news source)?

   b. What was viewed as the lowest risk? Are there any discernible patterns (e.g., younger/older, students v. faculty, news source)?

   c. What surprises resulted from the survey?

   d. How accurate are your results? (That is, are the results an accurate representation of the university population based on sample randomness, sample size, and time and location of sampling)?

    e. What could you do to improve the survey?

    f. If you were a public official for your town, based on the results of your survey,

       ○ What environmental problems would you focus on and why?

       ○ Are there other considerations besides risk?

       ○ Explain these other considerations.

    g. How do these rankings compare with other rankings or surveys in your city, state, or country? (Check on the Internet for comparative risk surveys and reference your source.)

Environmental Risk Survey

Hello. We are conducting a survey for a lab project. Completion of the survey will take approximately four minutes. We will keep your answers confidential and you will not be identified as a respondent. We will only report the total data for our use in this class.

Below is a list of environmental issues presented in alphabetical order. Please circle the number for each topic below that best shows your level of concern: 0 = no concern, 1=low concern, 2=medium concern, and 3=high concern

Environmental Issue	No	Low	Med	High
	0	1	2	3
	0	1	2	3
	0	1	2	3
	0	1	2	3
	0	1	2	3
	0	1	2	3
	0	1	2	3
	0	1	2	3
	0	1	2	3
	0	1	2	3

Please complete the information below as it applies to you:

Age: ___ <18 ___ 19–21 ___22–29 ___ 30–39 ___40–64 ___ 65+

Gender: ____ Male ____ Female

Status: ____ Part-time Student ____ Full-time Student ____ Staff ____ Professor

Primary source of news: ____ Television ____ Radio ____ Internet

                                             ____ Newspaper ____ Friends/family ____ Other: _____

Thank you for your participation.

Public Awareness and *Silent Spring*

OBJECTIVES

- Be able to describe Rachel Carson's role in environmental history and in public environmental awareness
- Be able to describe how social and governmental attitudes toward environmental toxicity changed in the late 20th century
- Be able to research and prepare brief biographies of key environmental figures

KEY CONCEPTS & TERMS

✓ DDT
✓ Environmental toxicology
✓ Rachel Carson

INTRODUCTION

Rachel Carson was the author of the seminal environmental book, *Silent Spring*. This book is credited as a catalyst in spawning the contemporary environmental movement in the 1960s. Her book focuses on society's seemingly over-reliance and careless application of synthetic pesticides, especially the class of pesticides known as organochlorines, such as **DDT**.

In this lab, you will be reviewing a documentary on Rachel Carson and her book, *Silent Spring*. This video is important for several reasons: Silent Spring shows what one person can do; it shows the importance of environmental awareness; and it is a reminder that these issues are on-going. *Silent Spring* also helped foster the field of **environmental toxicology**—the study of environmental poisons and their effect on humans and the environment. After viewing the documentary, be prepared to discuss your reactions and to do some research on other luminary figures in environmental science.

MATERIALS

☐ Video: **Rachel Carson's Silent Spring** (from PBS's "The American Experience")
☐ Access to the Internet or library

TASKS

Address the following questions as you watch this video. The answers will be discussed in class following the video.

1. When was *The Sea Around Us* written?

2. Was it her first book?

3. Her views were considered radical in the 1960s; do you think they would be radical in 1990? Would they be radical now? Why or why not?

4. In the late 1950s, were scientists asking how DDT killed insects?

5. Do you think the Bill of Rights should include the right to a healthy and safe environment? Who would define or enforce this? Explain.

6. What particular event or person inspired Rachel Carson to finally write *Silent Spring*?

7. Are there other chemicals currently in common use that could cause problems to the degree of DDT?

8. What was the government's attitude toward fire ants?

9. Was Rachel Carson against the use of pesticides? Explain.

10. Many scientists in the 1950s knew about the lethal nature of pesticides. What did Rachel Carson do that they did not do? What was her particular talent?

11. How did environmental awareness and public views of environmental toxicity change during the late 20th century?

Internet/Library research

12. Research another person who was important in environmental science or in the history of the environmental movement. Write a summary of who the person was and how that person's accomplishments or actions affected society.

13. Locate the most recent U.S. Environmental Protection Agency's *Pesticide Industry Sales and Usage*.[16] What are the major pesticides (by amount) and what are they used for? How much is spent to purchase pesticides? What observations can you make regarding pesticides use since 1980?

Class Discussion

- Discuss the use of scientific evidence in policy decisions on the use of pesticides.
- Discuss the video and reactions to it.
- Discuss the results of the summaries completed in item 12 above.

[16]http://www.epa.gov/oppbead1/pestsales/

CHAPTER TWENTY-SIX

Energy Conservation

OBJECTIVES

- Be able to identify basic energy conservation terminology
- Be able to conduct a basic building energy audit by gathering data and using formulae
- Be able to summarize and comment on energy usage in America

KEY CONCEPTS & TERMS

- ✓ British Thermal Units (BTUs)
- ✓ Energy audit
- ✓ Heat loss
- ✓ Kilowatt hours
- ✓ Lux
- ✓ R-value
- ✓ Wattage

INTRODUCTION

The United States consumes more energy than any other country. According to the U.S. Energy Information Administration, in 2001, the U.S. consumed 97.05 quadrillion BTUs in 2001.[17] In comparison, all the countries of South America, Central America, and Western Europe combined consumed 93.68 quadrillion BTUs, yet these regions had 624 million more people than the U.S. (EIA, 2004a).

Historically, the U.S. either has had access to abundant sources of energy, especially fossil fuels, or the economic ability and political determination to obtain inexpensive resources from other countries. As shown in Figure 26.1, the U.S. is still heavily dependent on fossil fuels as 90% of energy consumed is from coal, petroleum, and natural gas. Consequently, there has been minimal incentive in developing alternative forms of energy or in adopting a policy to consume energy more efficiently through an energy conservation program. The exception was during the Carter administration of the late 1970s where energy conservation measures—the 55 mph speed limit, solar energy tax credits, and recommended home thermostat settings—resulted from the OPEC oil embargo. We seem to

[17]A **British Thermal Unit**—BTU—is the amount of energy required to raise the temperature of 1 pound of water 1 degree Fahrenheit when the water is near 39.2° Fahrenheit.

have forgotten the long lines at gas stations in the 1970s as energy-saving measures have been largely ignored, reduced, or repealed.

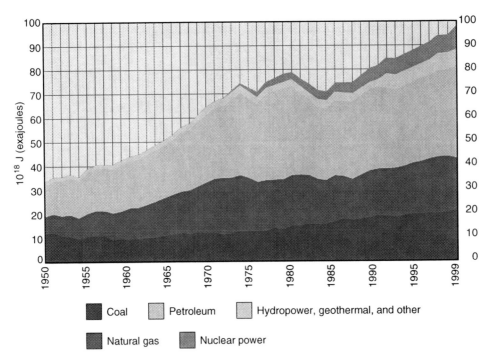

Figure 26.1—Total Energy Consumption in the U.S., 1950-1999.
Current increases seem to average about 5% per year. Source: Bodkin, D.B. and E. A. Keller. 2003. Environmental Science: Earth as a Living Planet. 4th ed. John Wiley & Sons, NY. (p. 313). Data from Energy Information Administration Annual Energy Review. EIA, Washington DC. (see http://www.eia.doe.gov/ for current energy information).

We have direct control over several categories of energy consumption in our homes, offices, and classrooms. We can conserve energy in our lighting, space heating, and cooling by using more efficient appliances, installing more efficient windows, landscaping to control external heating and cooling forces, and moderating our preferences.

MATERIALS

- ☐ Compass
- ☐ Yard sticks
- ☐ Inside/outside thermometer
- ☐ Liquid thermometer
- ☐ Calculator
- ☐ Light meter

- ☐ Watt meter (optional)
- ☐ Beaker
- ☐ 15-Watt incandescent lamp
- ☐ 15-Watt fluorescent lamp
- ☐ Scale

TASKS

In this lab, we will perform some basic components of an energy audit. (An **energy audit** is a survey that identifies how much energy is used as a means to identify measures to use energy more efficiently.) This lab requires you to calculate energy loss and translate this figure into economic cost. You also will conduct some basic comparisons with the data. Remember to show all your work, not just the final answer, and to write out the correct units for each number.

1. *Electrical Lighting*

Lighting is a significant source of energy consumption. Nationally, in the typical home, lighting accounts for about 5 to 10 percent of total energy use ($50 to $150). If extensive outdoor lighting is used, the annual lighting cost will be significantly higher. For this section, you will compare two types of light bulbs.

a. In a dark room hold a light meter approximately 4 feet from a 15-watt incandescent bulb and record the lux reading. The lux, symbolized lx, is the unit of luminance in the International System of Units. It is defined in terms of lumens per meter squared (lm/m^2).

b. Now measure the amount of light coming from a 15-watt fluorescent light bulb and record the lux reading. Hold the meter the same distance as the incandescent bulb and keep all other conditions the same.

c. Both bulbs have the same rated wattage; so theoretically, they consume the same amount of energy to produce light. (If you have a watt-meter, use it to check the watts consumed, as the actual current draw in watts may differ from the rated draw.) Which bulb is more efficient in providing light? You likely suspected this result, but now you can determine how much more efficient one bulb is over the other. Wattage is a measure of power input, and light output is measured in lux. To compare the efficiency, divide the amount of lux by the wattage for a lux per watt output.

d. Based on your measurements, approximately how much more efficient is this bulb (e.g., percentage difference)?

e. Assume a 20-watt fluorescent bulb produces approximately the same amount of light as a 70-watt incandescent bulb. What is the potential cost savings in using fluorescent bulbs? Multiply the wattage of each bulb by the number of hours it is used each year (assume 10 hours per day, every day) and divide by 1000 to get the kilowatt-hour (kWh) usage per year.[18] Multiply this by the current electricity price, which will give you the operating cost of that light for one year. Electricity prices vary widely across the U.S. In January 2004, according to the U.S. Energy Information Administration (EIA, 2004b), the national average price for residential electricity was 8.24 cents per kWh.

Fluorescent watts × hours/year/1000 = kWh/year

kWh/year × $/kWh = $/year

f. Note that you calculated the difference for only one light. Estimate the number of lights in an average three-bedroom house. Based on your estimate, what would the annual savings be if every incandescent light were replaced with a fluorescent light? What would the savings be in a city with 10,000 three-bedroom houses?

---

[18]The kilowatt-hour (symbolized kWh) is a unit of energy equivalent to one kilowatt (1 kW) of power expended for one hour. The kilowatt-hour is not a standard unit in any formal system, but is commonly used in electrical applications.

g. Provide your thoughts on energy conservation based upon the results of this investigation into electrical lighting.

2. *Space Heating*

A major source of waste energy is the loss of heat through windows. Loss of heat in cooler climates requires furnaces to work harder. According to the U.S. Department of Energy, in 1990 alone, the energy used to offset unwanted heat losses and gains through windows in residential and commercial buildings cost the U.S. $20 billion, which represented one-fourth of all the energy used for space heating and cooling (DOE, 2004). However, properly designed windows can significantly reduce this wasted energy.

Many windows are constructed with only a single pane, which has limited capacity to resist heat loss. A window's resistance to heat loss is referred to the R-value. (**R** is the resistance to heat flow in BTUs.) The R-value of a single-pane window is about 0.9. (An R-value of 1 is equal to the number of BTUs that would pass through a 1 ft² surface in 1 hour if the difference in temperature on opposite sides of the surface is 1° F.) Using the following formula (Q_{HOUR} = hourly heat loss in BTUs), you can calculate the hourly heat loss through a window:

$$Q_{HOUR} = \frac{\text{difference in temp (°F)} \times \text{ft}^2}{\text{R-value}}$$

a. In the lab, choose one single-pane window and measure its surface area in ft². If you do not have a single-pane window, check in other rooms and if you still do not find one, use the window you have.

b. Measure the <u>difference</u> in air temperature between the inside and outside of the window.

c. Calculate the <u>hourly</u> rate of heat transfer through the window by using the Q_{HOUR} formula.

d. Calculate the annual rate of heat transfer through the window. You will have to use a different formula because obviously the temperature difference fluctuates widely. Every building has a Building Load Coefficient (BLC) for heat loss, which is a function of the building design and the environmental conditions. We assume the building's design allows for adequate heating during the coldest day of the year, but we want to see on average what the heat loss is for a heating season or year (we assume no heat is needed in the summer months). The formula for annual heat loss, incorporating BLC, is:

$$Q_{YEAR} = (Q_{HOUR}/\Delta T)(24 \text{ hr/day})(DD)$$

ΔT represents the change in temperature between the average inside temperature and the coldest day of the heating season.

    Either use the temperature at the center of the lab or if you are doing this on a day in which the building requires no heat, use 65° F.

DD = heating degree-days

    DD are calculated as follows: DD = normal inside temp (65° F)—average 24 hour outside air temperature. These values are then summed for the entire year. Use the DD for the city nearest you, which you can determine though the National Weather Service Climate Prediction Center or other source.[19]

---

[19]A source for DD information is: http://www.weather2000.com/dd_glossary.html

e. What would happen if you turned the thermostat down during the winter? Calculate the effect of a 6° F decrease in the inside temperature on the hourly rate of heat transfer (Q_{HOUR}).

f. With double-pane windows, air trapped between the panes provides insulation. The R-value of a double-pane window is about 1.9. Calculate a revised Q_{HOUR} and then Q_{YEAR} if the single pane was replaced with a double-pane window.

g. Triple-pane windows are available, but because the extra pane of glass adds significantly to the weight and cost while only marginally improving the efficiency, they are not popular. Instead, there are super efficient double-pane windows. The air space is filled with an inert gas, generally argon, which has a heat transfer rate 30% lower than air. In addition, a special surface coating is used to reduce heat transfer through the window known as **low-emissivity (low-E) glass**. These coatings reflect 40% to 70% of the heat that is normally transmitted through clear glass, while allowing the full amount of light to pass through. An argon gas-filled, low-E double-pane window has an R-value of 3. Calculate a revised Q_{HOUR} and then Q_{YEAR} if the single pane was replaced with a gas-filled, low-E double pane window.

h. Energy loss is important data. However, to many, the more important figure is the energy loss in dollars. Calculate the energy savings potential by upgrading the window from single pane to a gas-filled, low-E double-pane window (R = 3) assuming the room is heated with oil. (You will need to calculate the cost for both, then subtract the two to determine the savings.)

cost of fuel oil($)/yr =

$$\frac{Q_{YEAR}}{(\text{BTU content of fuel oil})(\text{combustion efficiency})(\$ \text{ of fuel oil/gallon})}$$

BTU content of fuel oil = 145,000 BTUs per gallon

Combustion efficiency varies, but assume the combustion efficiency for the oil-fired boiler is 75%.

i. The cost for the previous calculation may appear low, but you calculated the savings from only one window. What is the cost in loss energy for all the windows in the room? (Assume all the windows are single-pane. Thus, you only need to change the surface area in your equation.)

j. The same thing we did for the windows could be done for the building's roof, doors, walls, and slab to determine an entire Q for the building. How do you think these other factors compare with the heat loss from the windows? What conclusions can you draw regarding windows and space heating based on your calculations?

3. Heating Water

According to the Department of Energy, 14% of the energy used in an average home is for heating water. Although this may appear to be minimal, there are over 100.5 million households in the U.S. and it all adds up. Leaky faucets are a significant source of wasted heated water.

a. Turn on a cold-water faucet so that it is leaking at a rate of about one drip per second. Collect this water in a beaker for 10 minutes. Determine the net weight of the collected water in pounds.

b. Assuming the faucet leaks at a constant rate 24 hours per day, calculate the amount of water that would be lost from this faucet in one year.

c. Measure the temperature of water fresh from a cold-water faucet in degrees Fahrenheit. Assume the water entering the water heater is the same as the temperature of the water coming out of the faucet. Assume it leaves the heater at 140° F (a common setting); the difference between the two temperatures = ΔT_{H2O}.

d. Assume you have a hot-water faucet that drips at the rate you calculated above for the cold-water faucet. Determine the BTUs of lost heat energy in one year if the hot-water faucet dripped at that rate constantly. (It takes 1 BTU of heat energy to raise the temperature of 1 pound of water 1° F.) BTUs/year = (lb water in 10 min/hr) (6 min × 24 hrs/day)(365 days/year)(ΔT_{H2O})

e. Many hot water heaters use electricity to heat the water. How many kilowatt-hours of electricity are used to heat the yearly loss of water through the leaky faucet? (There are 3,412 BTUs in a kilowatt—kWh—of electricity.)

f. What is the cost in dollars for this water loss.

g. Note that you calculated the energy loss from only one faucet. Other than the price of the electricity used to heat the water, what factors should we have considered when confronted with a legion of leaky faucets? What conclusions can you draw regarding energy use and leaky faucets based on your calculations?

4. Landscaping for Energy Conservation

Factors outside as well as inside affect the energy consumption of a structure.[20] Landscaping for energy conservation can reduce household energy costs by providing screening from the hot sun in summer and blocking cold wind in winter as depicted in Figure 26.2. About one-fourth of a household's energy consumption for heating and cooling can be eliminated by careful landscape design. Computer models developed by the U.S. Department of Energy indicate the average household can save between $100 and $250 in annual energy costs through proper placement of a few trees.

An 8-foot (2.4-meter) deciduous tree, for example, can save a household hundreds of dollars in reduced cooling costs, yet still admit some winter sunshine to reduce heating and lighting costs.

Summer

Have you noticed that parks and wooded areas seem cooler than nearby city streets? It is true; shading and transpiration from trees can reduce surrounding air temperatures as

[20]The information in this section comes from the U.S. Department of Energy fact sheet on landscaping for energy efficiency: http://www.eere.energy.gov/erec/factsheets/landscape.html

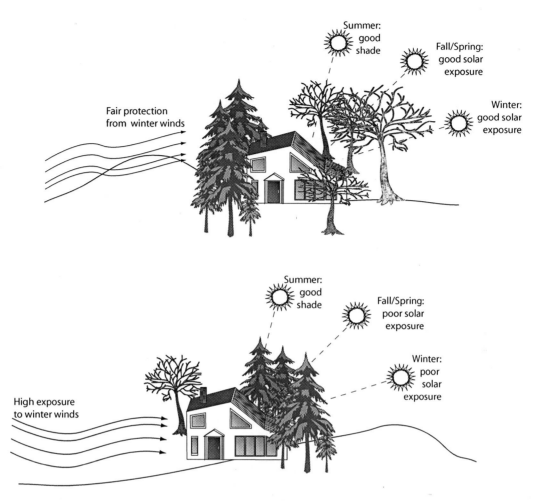

Figure 26.2—Effective Landscaping for Energy Conservation.

much as 9° F (5° C). A well-planned landscape can reduce a non-shaded home's summer air-conditioning costs by 15% to 50%.

Winter

Wind chill affects us because the wind whisks away warmth. A 20 mph (32 kmh) wind in an outside temperature of 10° F (−2° C) results in a below-zero wind chill of −9° F (−12.8° C). Good design can drastically cut down on wind chill. Trees, fences, structures, and landscape forms (e.g., slopes, **berms**, etc.) can help shield your house from the wind. However, in cold climates, windbreaks should not block the winter sun from reaching southern windows.

    a. Walk completely around the perimeter of your classroom building. For each side (e.g., west, south, etc.) how many deciduous trees, coniferous tress, shrubs, buildings, berms, and other features exist that could serve as a windbreak?

    b. For each side, estimate the percent of the building's surface <u>not</u> shaded at least a part of the normal work hours.

    c. Inventory each tree, feature, building, and so on, that provides shade. Determine whether this shade provides cooling during the summer months when it is needed or provides cooling during winter when it is not desirable. Note any assumptions you used in making this estimate. Do the landscape and setting provide for efficient heating and cooling? What changes would you suggest?

d. Evaluate the design and orientation of the building itself for energy efficiency and solar gain. Summarize your evaluation. Does the building's design allow for efficient heating and cooling? What changes would you suggest?

5. Energy Conservation—The Big Picture

The "big picture" happens one building at a time.

a. Based on your answers to Sections 1, 2, 3, and 4, what conclusion can you make regarding wasted energy and energy conservation in your classroom?

b. You examined how energy is wasted and calculated the amount. But there is more to the problem. Do some research and summarize the relationships between waste energy and the following: acid precipitation, airborne mercury, and global climate change.

c. Can energy conservation be improved? If so, what should be the focus—list some specific recommendations. Try the U.S. Department of Energy's home audit for your home: http://www.homeenergysaver.lbl.gov. What did you find out?

References

U.S. Department of Energy (DOE). 2004. Consumer Energy Information: EREC Fact Sheets. [Online] Available at http://www.eere.energy.gov/erec/factsheets/eewindows.html (verified 8 June 2004).

U.S. Department of Energy Fact Sheet on Landscaping for Energy Efficiency. [Online] Available at http://www.eere.energy.gov/erec/factsheets/landscape.html (verified 8 June 2004).

U.S. Energy Information Administration (EIA). 2004a. International Energy Annual 2002. Report Released: March-May 2004. [Online] Available at http://www.eia.doe.gov/pub/international/iealf/tablee1.xls (verified 8 June 2004).

U.S. Energy Information Administration (EIA). 2004b. Average Retail Price of Electricity to Ultimate Customers by End-Use Sector, by State, January 2004 and 2003. [Online] Available at http://www.eia.doe.gov/cneaf/electricity/epm/table5_6_a.html (verified 8 June 2004).

PART THREE

Applied Problem Sets

ONE

The Scientific Method: Observation and Hypothesis

INTRODUCTION

By now, you have examined a variety of phenomena in the natural world. A focus of environmental science is to explain these phenomena. That is, what are the causes and effects of such phenomena? How does one go about investigating the causes and effects such that the conclusions are meaningful and reliable? The answer is the **scientific method**, which uses designed experiments and careful observations to investigate causes and effects.

The scientific method is the systematic procedure for investigation, which generally is composed of the following:

1. Observe a phenomenon

2. Formulate a hypothesis: a tentative description that explains your observation

3. Design and conduct an experiment to test the hypothesis

4. Present your results

5. Interpret the results that validate or modify the hypothesis

Step 1—Observe a Phenomenon

Suppose you observe that an environmental phenomenon (e.g., rainbow trout, *Salmo gairdneri* are no longer present in a nearby polluted river). You could conduct research by examining the literature to determine what other studies have been done to help you find an explanation. However, the studies may not be directly applicable. You might want to design and conduct your own study.

Step 2—Formulate a Hypothesis

The second step is to formulate a hypothesis. A **hypothesis** is a statement of the cause and effect in a specific situation. It is a tentative statement that proposes a possible explanation of the observed phenomenon. A hypothesis should be testable, unambiguous, and dichotomous (i.e., a yes or no statement). The key word is *testable* because you will need to design and perform an experiment on how the two variables might be related.

In developing an hypothesis, you must form a set of two contradicting hypotheses: the null hypothesis (symbolized by H_0) and the alternate hypothesis (symbolized by H_A). Basically, the null hypothesis is a stated assumption that there is *no* effect in a cause-and-effect process or relationship. In contrast, the alternate hypothesis is a statement that there is an effect.

Your hypotheses can be as general or as specific as needed. However, a very specific hypothesis is much easier to test in a meaningful way that can yield useful data.

Example of general hypotheses:

H_0 Dissolved oxygen does not affect fish
H_A Dissolved oxygen affects fish

As you can see, the hypothesis is very general and is not very meaningful. What type of fish? What is meant by affect? How much dissolved oxygen? How can it be tested?

Example of specific hypotheses:

H_0 Less than 3 mg/l of dissolved oxygen is not lethal to juvenile Rainbow trout (*Salmo gairdneri*)
H_A Less than 3 mg/l of dissolved oxygen is lethal to juvenile Rainbow trout (*Salmo gairdneri*)

These hypotheses are testable as you know exactly what to test and the results are meaningful in relation to the hypothesis. They are also unambiguous and dichotomous.

Step 3—Design an Experiment

The next step is to design an experiment to test your hypothesis. Your experiment is affected by variables. There are three kinds of variables in an experiment: independent, dependent, and controlled. The **independent variable** is the variable you purposely manipulate. The **dependent variable** is the variable that is being observed, which changes in response to the independent variable. Variables that are not changed are called **controlled variables**.

When conducting an experiment, it must be a controlled experiment. That is, you must compare an "experimental or treatment group" with a "control group." The two groups are <u>exactly</u> the same except for the one variable being tested. For example, in your experiment on dissolved oxygen and trout, you may select to use two aquariums and 5 juvenile rainbow trout. The aquarium, food, light source, location of the aquarium, ambient temperature, water temperature, trout stock, trout gender, and age of trout all have to be exactly the same. However, the water in one aquarium has a dissolved oxygen level of 3 mg/l and the other has a "normal" level. The aquarium with "normal" dissolved oxygen is called the **control** and the aquarium with the lowered dissolved oxygen is called the **treatment**. The control group acts as a reference point to compare with the treatment group. If the test is properly designed and constructed, any difference between the two groups can only be due to the one experimental factor tested. The challenge is to control for variables that may affect (confound) your results. A **confounding variable** is an extraneous (uncontrolled) variable that could produce an alternative explanation for the results. Again, controlling confounding variables is done by ensuring that the two aquarium tests are the same except for the treatment; in this case the dissolved oxygen.

In conducting your experiment, how can you be certain of the results? Was it by chance that the experiment ended the way it did? The key component is **replication**. Your experiment should be repeated several times to reduce the likelihood that the results are by chance. For example, in your dissolved oxygen experiment, you should use multiple fish in each aquarium <u>and</u> repeat the same experiment at least two additional times.

EXERCISES

A. Hypothesis Formulation

1. Research the meaning of *null hypothesis*. Describe how and why it is used in experimental design. Properly cite your reference.

2. The following are observations of environmental phenomena. Rewrite each observation into a formal, testable null hypothesis. Find a formal, testable alternate hypothesis. Put the dependent variable in **bold** font, and underline the independent variable for each.

   a. Eggshells exposed to mercury seem to be thinner.

   b. Plants near roads where salt is used in the winter appear smaller.

   c. Gulls covered with oil do not look like they can fly.

   d. There seems like there are fewer squirrels in January.

   e. Trees in areas where there is acid precipitation look shorter.

3. Describe five environmental phenomena affecting your region.

4. For each of these environmental phenomena, create formal, testable hypotheses (null and alternate hypothesis). **Bold** the dependent variable and underline the independent variable for each.

B. Experimental Design

5. From question 4:

   a. Choose one of the phenomena and its hypotheses (H_0 and H_A).

   b. Use lines and arrows to outline and connect the steps (this is a "block-flow diagram") in your experimental design to test the hypothesis based on the scientific method.

   c. List the materials and equipment needed for your experimental design. Provide enough details to indicate how the test would work.

6. Based on question 5, write a critique of your experimental design. Your critique should include (a) likely confounding variables, (b) how or what could be done to improve the design, and (c) how to refine the hypothesis.

TWO

The Scientific Method: Results and Discussion

INTRODUCTION

In *Problem Set 1* you studied the basics of the scientific method, the systematic procedure for investigation, which generally is composed of the following:

1. Observe a phenomenon

2. Formulate a hypothesis (H_0 and H_A): a tentative description that explains your observation

3. Design and conduct an experiment to test the hypothesis

4. Present your results

5. Interpret the results that validate or modify the hypothesis

You also conducted exercises to formulate a proper, testable hypothesis and to design an experiment to test your hypothesis. Following the completion of your experiment you will have generated data. In this homework, you will focus on steps 4 and 5 of the scientific method, *presenting your results* and *interpreting your results*.

Step 4—Present Your Results

The data obtained from your experiment are presented in the results section. Because the data are generally numbers, the data are presented in tables or figures. Note that in the results section, you do <u>not</u> interpret the results. This is done in the discussion section.

The preliminary data collected from your experiment is referred to as **raw data**. Raw data is data that has not been processed (extracted, organized, summarized, or formatted) for presentation. You thus need to summarize and organize your data before presenting it. For example, you may want to present basic, descriptive statistics such as mean or mode, or percentages. Consequently, you will need to perform some basic calculations.

Data is presented in tables as numbers and symbolically in charts and graphs, which are used to depict trends. Tables and figures can be produced in word processing, spreadsheets, or statistics programs. As presented below, the major data presentation formats are the table, histogram, line graph, and pie chart. Note that titles are placed on top of tables and placed below figures (histograms, line charts, and pie charts).

Table

Tables are used to present data in a tabular or spreadsheet format. See Table P2.1 for an example.

TABLE P2.1 Example of a Data Table for Testing a Hypothesis	
Day	**Night**
37	56
42	22
15	13
13	24
11	16

Histogram

The histogram is a bar chart that depicts frequencies. Generally, the independent variable is plotted along the X (horizontal) axis and the dependent variable is plotted along the Y (vertical) axis. The independent variable can attain a finite number of discrete values (e.g., 10) rather than a continuous range of values. For continuous values, the line graph is used. See Figure P2.1 for an example of a histogram.

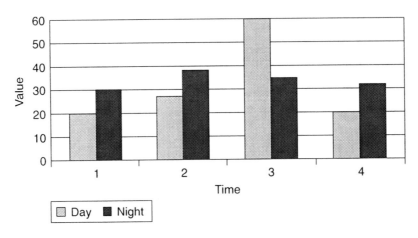

Figure P2.1—Example of a Histogram.

Line Graph

A line graph is used to depict the trend of one or more items over a period of time or number of events. Figure P2.2 is an example of a line graph.

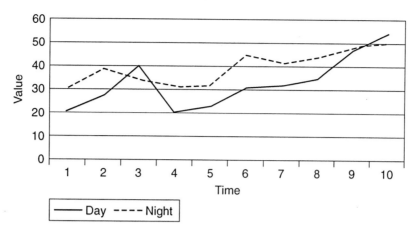

Figure P2.2—Example of a Line Graph.

Pie Chart

A pie chart is a circle divided into segments with each piece of the pie representing some data. Pie charts are generally used to depict percentage or proportional composition. Figure P2.3 is an example of a pie chart. Be sure to give the units for all data on the chart.

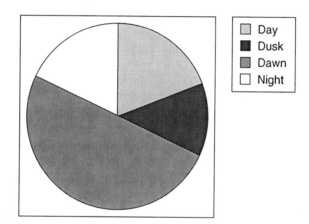

Figure P2.3—Example of a Pie Chart.

Step 5—Interpret the Results

The final step in the scientific method is to interpret and discuss your results. At this step, you examine the results and state whether your hypothesis is accepted or rejected. If your hypothesis is accepted, briefly explain the major factors that support it. If, however, your hypothesis was not accepted, try to explain your results: Was there a flaw in your experimental design? Were there confounding variables that may have impacted your results? An important aspect of the scientific method is that rejecting your hypothesis is perfectly acceptable because the purpose of the scientific method is to learn something from the process. If you reject your hypothesis, you can refine your hypothesis or experimental design and retest.

To support your hypothesis you must ask the question, are the results significant? Are there tests I might use to tell me if the results are significant?

Notice that the word "prove" was not used. In science, you do not prove anything because science is based on **induction**—a type of reasoning where one uses specific examples or facts to make generalities or hypotheses). The inductive method entails drawing generalized conclusions based on evidence. Thus, scientists can only draw conclusions on what they find, not on what they cannot find. In other words, there is always the untested hypothesis. This does not mean scientists lack confidence in findings, but that there is recognition that an alternative explanation may always be possible, even though it may be a very slim chance. Instead of using the word *prove*, scientists tend to use words like the following: supports, agrees with, suggests, affirms, upholds, and demonstrates.

EXERCISE

A. *Presenting Results*

1. Create a histogram using Excel or other suitable computer program using the following hypothetical data. The figure should report the amount (in metric tons) of hazardous waste produced by each state in 2004. Be sure to label the figure with an appropriate title and clearly identify the name and units for each axis.

   Alabama 423,968; Alaska 4,547; Arizona 53,031; Arkansas 1,052,744; California 672,946; Colorado 82,021; Connecticut 60,219; Delaware 19,353; and Florida 398,535.

2. With the above data, create a pie chart.

3. Create a table with the following hypothetical data: In 1999, 53 snowshoe hare (*Lepus americanus*) were trapped in study area A (this area has 1 bobcat per 10 km^2). In 1998, there were 51; in 1997, 48 were trapped; and 43 in 1996. In 2000, the number trapped was 56 and 63 in 2001. This is in contrast to study area B (this area has 1 bobcat per 15 km^2): the number trapped in 1998 was 84; in 1996, 79 (the same as in 1999); in 1997, 81; in 2000, 78; and in 2001, 80.

4. In accordance with the Emergency Planning and Community Right-to-Know Act, specified industries (e.g., companies that manufacture or process 25,000 lbs per year or otherwise uses 10,000 lb per year of some 600 designated toxic chemicals) must report releases of these designated toxic chemicals during the previous calendar year. According to the reports submitted to the U.S. EPA, facilities have reported the release of chlorine into the air and water since 1988 (in pounds). Convert this data into kilograms and place this data into an appropriate graph. Insert commas in the appropriate places—this makes the numbers easier to read on your graph.

YEAR	AIR (LB)	SURFACE WATER (LB)
1988	1334786	369118
1989	11378265	231874
1990	10239577	397270
1991	9652721	467719
1992	9060110	455128
1993	6200514	543651
1994	5135964	373696
1995	4837376	224936
1996	4001439	144822
1997	3716432	584709
1998	3939826	697038
1999	3852954	495955
2000	3954012	412062

B. Exercise—Discussion

5. Describe any discernible pattern in the air data and the surface water data from question 4. Explain what factors may contribute to a difference.

6. Your hypothesis was that larger states generate more hazardous waste than smaller states. Based on the results presented in question 1:

   a. Can you support this hypothesis?

   b. What are some potentially confounding variables?

   c. What could you do to further increase your confidence in your results?

7. Your hypothesis was that snowshoe hare trapped in study area A would be at a lower rate than in study area B. Based on the results presented in question 3:

   a. Can you support this hypothesis?

   b. What are some potentially confounding variables?

   c. What could you do to further increase your confidence in your hypothesis?

8. Why is it important to be able to interpret environmental information using graphs?

PROBLEM SET

THREE

Quantification of Environmental Problems

In this problem set, we will explore two fundamental components of expressing and quantifying environmental problems: scientific notation and calculating percentage changes over times.

I—SCIENTIFIC NOTATION

Environmental science often deals with very large numbers (e.g., biomass production of a forest) and very small numbers (contamination of an aquifer). To better manage these data, and to decrease errors, scientists have developed a shorter method to express numbers using *scientific notation*. Scientific notation is based on powers of the base number 10. Thus, for example, an environmental scientist calculates that 146,000,000,000 kilograms of biomass was produced in a test plot during the previous year. Using scientific notation, the number would be:

$$1.46 \times 10^{11}$$

To write a number in scientific notation for 146,000,000,000:
Step 1: place the decimal after the first digit and drop all the zeroes.

$$1.46 \qquad 000,000,000$$

Step 2: count the number of places from the decimal to the end of the number. For example:

1	.	4	6	0	0	0	0	0	0	0	0	0
Count	.	1	2	3	4	5	6	7	8	9	10	11

There are 11 places after the decimal point; therefore the exponent is **11**. Thus, 146,000,000,000 becomes 1.46×10^{11}.

$$152,000,000 \text{ is } 1.52 \times 10^{8}$$

Exponents in scientific notation are often expressed in different ways. For example, 146,000,000,000 can also be written as:

1.46E + 11 or as 1.46 × 10^11

A similar approach is used for small numbers (<1), which will have a negative exponent. A millionth of a second is:

0.000001 sec. = 1.0×10^{-6} (or 1.0E −6 or 1.0^ −6)

Thus, with small numbers, you count from the decimal the number of zeroes until you reach the first non-zero number.

0.00000123 sec. = 1.23×10^{-6}

Exercise: Scientific Notation

Convert the following numbers into or from scientific notation:

1. 678,950,000,000 = _____

2. 1,000,000,000,000 = _____

3. .00000004567 = _____

4. 5,689,000,000 = _____

5. 8923 = _____

6. .000000045678963 = _____

7. 2.223×10^{-9} = _____

8. 3.19E + 12 = _____

9. $4.444^{\wedge} -4$ = _____

10. 7.47×10^{13} = _____

II—CALCULATING PERCENT CHANGE

Environmental scientists often analyze trends. A popular approach to communicating these trends is percentage increase/decrease over time. In calculating percentage of change (whether an increase or a decrease), you are concerned with the difference between two numbers and how much of the first number added to or subtracted from the first number will produce the second number.

The three steps to calculate a percentage are:

1. Subtract

2. Divide

3. Multiply

Subtract: New number − original number = change
Divide: by original number
Multiply: by 100

Thus, the basic formula is simple: $\dfrac{\text{Change}}{\text{Original number}} \times 100 = \_\_ \%$

Percent increases and decreases are calculated with respect to the value <u>before</u> the change took place—the original number.

A. In 2004, the level of acetone in the river was 100 parts per million (ppm). In 2004, the level increased by 12 ppm. What is the percentage increase in acetone?

$$12/100 = .12 \times 100 = 12\%$$

B. In 2000, the level of ammonia was 63 ppm. In 2004, the level is 112 ppm. What is the percentage of increase since 2000?

Step 1: Determine the amount of change.

$$112 - 63 = 49$$

Step 2: Divide the change by the original number, then multiply by 100.

$$49/63 = .777 \times 100 = 77.7\%$$

C. Last month, the level of hexachlorobenzene had decreased by 8 ppm. If the level is now 47 ppm, by what percentage did the level of hexachlorobenzene decrease? The amount of change is 8, therefore the original amount is 47 + 8 = 55.

$$8/55 = .145 \times 100 = 14.5\%$$

D. A utility's operating costs for its electrostatic precipitator was the following:

2001 = $345,000
2002 = $325,000
2003 = $345,000

What were the annual percentage changes?

2001 to 2002: ($325,000 − $345,000)/$345,000 × 100 = −5.79%
2002 to 2003: ($345,000 − $325,000)/$325,000 × 100 = 6.15%

Exercise: *Calculating Percentage Changes*

Solve the following.

1. 234.98 to 324.77 = _____% change

2. 324.77 to 234.98 = _____% change

3. 7.14×10^7 to 8.47×10^8 = _____% change

4. 100 increases by 300% = _____

5. 756 declines by 100% = _____

6. 756 increases by 100% = _____

7. A utility's <u>total</u> pollution control costs are broken down as follows:

   Air pollution control = $234,000
   Wastewater treatment = $167,000
   Solid waste = $45,000
   Hazardous waste = $12,000

   What percentage does the company spend for each?

   Air pollution control = _____%
   Wastewater treatment = _____%
   Solid waste = _____%
   Hazardous waste = _____%

8. The company's managers want to budget sufficient money for next year. Assuming that costs will increase by 2.4%, what will the costs be for next year for each category?

   Air pollution control = $ _____
   Wastewater treatment = $ _____
   Solid waste = $ _____
   Hazardous waste = $ _____

9. An environmental scientist has the following data on mercury concentrations in small mouth bass (*Micropterus dolomieu*) sampled from a pond. What is the annual mean concentration for each year? What is the mean for all three years?

2002	2003	2004
7.2×10^4	2.1×10^4	$6.4 \times F10^4$
6.5×10^4	4.1×10^4	7.7×10^4
7.1×10^4	7.0×10^4	5.2×10^4
6.6×10^4	7.3×10^4	7.2×10^4
6.9×10^4	5.9×10^4	7.1×10^4
7.1×10^4	5.0×10^4	9.4×10^4
7.5×10^4	5.1×10^4	2.1×10^4

10. What is the percentage difference between 2002 and 2003 and between 2003 and 2004 based on the annual, average mercury concentration?

11. Why is it important for a member of an environmentally literate society to be able to understand scientific notation and percentages?

Ecosystem Diagram

Botkin and Keller (2003:43) define ecosystem as "a community of organisms and its local nonliving environment in which matter (chemical elements) cycles and energy flows." Sustained life on earth depends on ecosystems, not on individual species or populations. Sometimes the boundaries of an ecosystem are well defined, and sometimes they are vague. An ecosystem may be a large forest or a tiny puddle; whatever size, it must have the flow of energy and the cycling of chemical elements. An ecosystem can be artificial or natural, or a combination.

Your task is to provide an illustration that could be used in a high school environmental science textbook. Select a local ecosystem that is particularly familiar to you. Draw the ecosystem. Some things to consider: feedback arrows (positive and negative), energy paths, biogeochemical cycles, habitats, dominant species, biodiversity, complexity, and organism interrelationships (i.e., predator and prey). This is not an art class, but your drawing does need to be clear and easy to understand and contain a significant amount of factual information.

After you draw the ecosystem, answer the following:

1. Where is this ecosystem found?

2. What roles are filled by this ecosystem in terms of the surrounding environment?

3. What are the principal characteristics of this ecosystem?

4. What is necessary to sustain this ecosystem?

Reference

Botkin, D.B. and E.A. Keller. 2003. Environmental Science: Earth as a Living Planet. 4th ed. Wiley & Sons, NY.

Biogeochemical Concept Diagram

Recall the examples of systems and biogeochemical cycles in the textbook and handouts. For this homework you will create a concept map diagram. A concept map is a way to visually understand complex information; essentially, it is a diagram that explores the relationships in knowledge and information within a topic.[21] The concept map contains nodes or cells, each with a concept, item, or question. The nodes or cells are linked, with the links labeled and direction of flow indicated with an arrow symbol. The labels explain the relationship between the nodes. The arrow reads like a sentence to describe the direction of the relationship.

Create your own concept map for **one** of the following "big six" (macronutrients): carbon, hydrogen, oxygen, nitrogen, phosphorus, and sulfur. Be sure your concept map addresses time, chemical reactions, pathways, and other appropriate factors. Figure P5.1 is a sample concept map for the hydrologic (water) cycle.

[21]As of the writing of this text (2004), this site contains free downloadable software for making concept maps: http://cmaptoolsv3beta.coginst.uwf.edu/. Other software packages are available—you can get a trial version of SmartDraw at http://www.smartdraw.com/specials/flowchart.asp, or concept maps can be made by hand as a group consensus or individual activity.

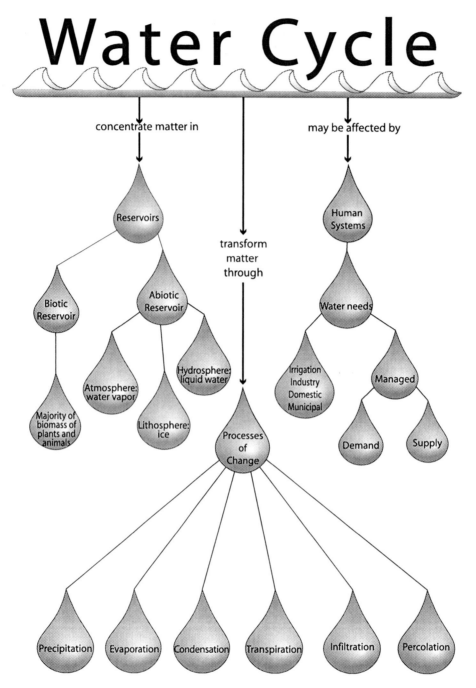

Figure P5.1.—Sample Concept Map—The Hydrologic Cycle.

Global Warming, CO$_2$, and You

INTRODUCTION

The economies of the industrialized world are dependent on fossil fuel. Coal, gas, and petroleum, formed hundreds of millions of years ago by decaying plants and animals, have provided modern people with a supply of stored energy from the sun. Fossil fuels have allowed us to move from a society based primarily on energy from people and living plants and animals to one based on fossil fuels. Special conditions that existed when coal, gas, and petroleum formed are not present now, so they can no longer form in significant amounts, if at all. Furthermore, formation of fossil fuels is a very slow process, too slow for replacement to keep step with current use.

Limited supplies are, however, not the only concerns. When fossil fuels are burned they produce, among other pollutants, carbon dioxide, the principal contributor to the greenhouse effect. About three-quarters of the anthropogenic (human-produced) emissions of **carbon dioxide** (CO$_2$) to the atmosphere during the past 20 years is due to fossil fuel burning. The rest is predominantly due to land-use change, especially deforestation (IPCC, 2001). The average annual release of carbon from fossil fuels during the 1990s was 6.35 billion tons (Marland et al., 2003). As shown in Figure P6.1, the ambient concentration of CO$_2$ in the atmosphere has increased about 29% since the beginning of the industrial revolution. The significance of this dramatic increase in ambient CO$_2$ is global warming (climate change) as depicted in Figure P6.2 on the **greenhouse effect**. Many scientists believe that stabilizing the climate will require slashing worldwide CO$_2$ emissions in half. Because the planet's population is now more than 6 billion people, each person's rightful share of CO$_2$ emissions is about 1.3 metric tons annually. Experts estimate that Americans generate 20 metric tons of CO$_2$ per capita each year, but the figure is misleading because it lumps together government, industrial, corporate, and personal production of CO$_2$.

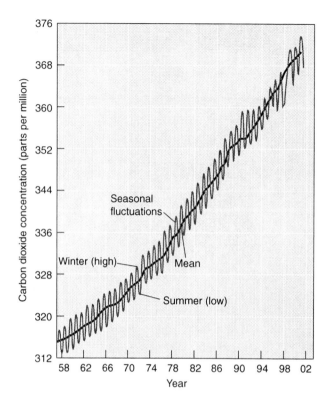

Figure P6.1—Atmospheric Concentration of Carbon Dioxide (CO₂) from 1958 to 2001.

This graph is based on data from the Mauna Loa Observatory in Hawaii—an ideal facility for measuring undisturbed air—constituting the longest continuous record of atmospheric carbon dioxide (CO_2) concentrations available in the world. Note the steady increase of CO_2. The seasonal fluctuations correspond to winter (a high level of CO_2) when plants are not actively growing and absorbing CO_2, and summer (low CO_2), when plants are more active. In 2002, the CO_2 concentration increased to an average of 373.1 parts per million by volume (ppmv). Sources: Raven, P.H. and L.R. Berg. 2004. Environment. 4th ed. John Wiley & Sons, NY (p. 462) and Keeling, C.D. and T.P. Whorf. 2004. Atmospheric CO_2 records from sites in the SIO air sampling network. In Trends: A Compendium of Data on Global Change. Carbon Dioxide Information Analysis Center, Oak Ridge National Laboratory, U.S. Department of Energy, Oak Ridge, TN. Data from Carbon Dioxide Research Group, Scripps Institution of Oceanography, University of California (http://sio.ucsd.edu/).

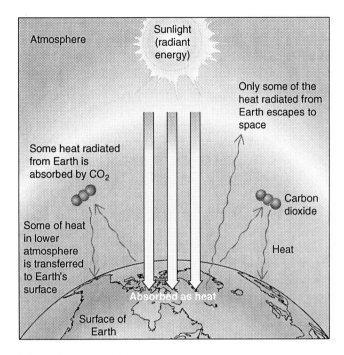

Figure P6.2—Enhanced Greenhouse Effect.
Carbon dioxide (CO_2) and other greenhouse gases accumulate, warming the atmosphere by absorbing some of the outgoing infrared (heat) radiation. Some of this heat is transferred back to Earth's surface, warming the land and ocean. Source: Raven, P.H. and L.R. Berg. 2004. Environment. 4th ed. John Wiley & Sons, NY (p. 463).

How much CO_2 do you contribute through your personal daily activities? To make a rough calculation of your emissions, follow the steps below. Be sure to keep track of your units.

1. Estimate the number of miles you drive per year and the average miles per gallon your car gets.

    _____miles/year divided by _____ miles/gallon = _____gallons/year

Burning a gallon of gas produces 9 kilograms of CO_2, so multiply the above total by 9 to get your total CO_2 emissions from driving.

    _____gallons/year multiplied by 9 kg/gallon = _____kg/year

2. Estimate the number of kilowatt-hours of electricity you used last year. Multiply the number on your last electric bill by 12 to get an estimate for the year. If you don't have an electric bill use the "Energy Requirements of Household Appliances" table and add the values for each appliance to find your annual energy consumption for electrical appliances in kilowatt-hours per year. One kilowatt-hour of electricity generated in a coal-fired power plant produces about 1 kg of CO_2. Nuclear energy is effectively free of CO_2, but produces only 20% of the nation's electricity. Assume 80% of your electricity comes from a coal-fired power plant.

    _____kilowatt-hours/year multiplied by 0.8 = _____kg/year

3. Estimate the number of BTUs of natural gas you used last year. Multiply the number on your last monthly gas bill by 12 to get an estimate for the year. If you don't have a gas bill, assume a typical value of 60,000,000 BTUs/year. 100,000 BTUs of natural gas produce 5.5 kg of CO_2. 100,000 BTUs is approximately the yield of 100 cubic feet of gas. 100 cubic feet of gas = 1 ccf (also called 1 Therm).

Using your gas bill or the typical value, determine how many kg of CO_2 is emitted per year.

4. Estimate the number of gallons of oil you used last year. If you don't have an oil bill, assume a typical value of 600 gal/yr. One gallon of oil generates 7.6 kg CO_2.

   _____gal/year multiplied by 7.6 kg CO_2/gal = _____ kg/year

5. If your home is heated by coal or wood assume, 100 lb of coal = 1,111,100 BTUs and 120 lb of dry wood = 948,000 BTUs. One cord of wood emits 3,068 kg of CO_2. However, since wood is only a temporary store of CO_2, some people argue that heating with wood does not make a contribution to atmospheric CO_2. What is your opinion?

6. Estimate the number of miles you flew last year. Airplane fuel efficiency varies, but on average flying one mile generates 0.23 kg (0.5 pound of CO_2) per passenger.

   Number of miles flown_____miles/year multiplied by 0.23 kg/mile = _____kg/year

7. Add the totals of the above calculations.

   _____kg/yr (car) + _____kg/yr (electricity) + _____kg/yr (natural gas) + _____kg/yr (oil)
   + _____kg/yr (wood, coal) + _____kg/yr (airplane) =_____kg/yr TOTAL

8. The figure you calculated is energy you use directly. It does not include the energy you use indirectly. Indirect energy use includes energy used in manufacturing products you buy, growing and processing food, and transporting food and products to you. Approximately 75% of the energy we use is used indirectly. Therefore you need to multiply the total kg/yr you calculated by 4 to obtain your total energy

   _____kg/year (direct) multiplied by 4 = _____kg/year (direct and indirect)

9. Now convert to metric tons.

   _____kg/year total divided by 1000 kg/metric ton = _____metric ton/year

   a. How does your total compare to the value of 1.3 metric tons that is estimated to be your rightful share?

   b. How does your total compare to the approximately average production of CO_2 per person in the U.S. of 20 metric tons?

   c. List some concrete things you do to reduce your CO_2 emissions and your contribution to global warming.

10. Using the Internet or library resources, what controversies exist concerning global warming?

TABLE P6.3	Energy Requirements of Household Electrical Appliances (kilowatt-hours consumed annually)

Air conditioner—860 (based on 1000 hours yr)	Humidifier—163
Electric blanket—147	Iron—60
Blender—1	Microwave—100
Broiler—85	Mixer—2
Clock—17	Radio—86
Clothes dryer—993	Shaver—0.5
Radio + CD player—109	Toaster—39
Clothes washer—103	Range + oven—596
Coffee maker—140	Hair dryer—25
Refrigerator w/defrost—1591	Window fan—200
Dehumidifier—377	Water heater—4219
Dishwasher—165	TV—320
Sewing machine—11	Fan (furnace)—650
Fan (circulating)—43	Hot plate—90
Freezer, 16.5 ft³, automatic defrost—1820	Heating pad—10
Vacuum cleaner—46	Waffle iron—20
Frying pan—100	Garbage disposal—7

Sources: Department of Energy, 2004.

WAYS TO REDUCE CARBON EMISSIONS

What you can do	*Reduction
Tune up your car	120 kg
Drive a car with 30 mpg instead of 20 mpg	330 kg
Drive a car with 40 mpg	599 kg
Drive a car with 50 mpg	717 kg
Take a train rather than fly	4.5 kg/160 km

*Estimation based on driving 10,000 miles/year

WAYS TO SAVE ENERGY

What you can do	Energy Savings
Improve insulation in your hot water heater	300 kW/yr
Switch from typical refrigerator/freezer to more energy-efficient model	2000 kW/yr
Substitute 18-watt compact fluorescent bulb for a regular 75-watt bulb for 8 hrs	170 kW/yr

*Every kilowatt-hour saved reduces carbon emissions by 0.2 kg.

References

Department of Energy (DOE). 2004. Office of Energy Efficiency and Renewable Energy. Consumer information [Online]. Available at http://www.eere.energy.gov/consumer-info/refbriefs/ec7.html (verified 9 June 2004).

Intergovernmental Panel on Climate Change (IPCC). 2001. Climate Change 2001: The Scientific Basis—Summary for Policymakers

Lyman, Francesca. 2001. The Greenhouse Trap by World Resources Institute: What We're Doing to the Atmosphere and How We Can Slow Global Warming. A World Resources Institute Guide to the Environment. WRI, Beacon Press.

Marland, G., T.A. Boden, and R.J. Andres. 2003. Global, Regional, and National CO_2 Emissions. In Trends: A Compendium of Data on Global Change. Carbon Dioxide Information Analysis Center, Oak Ridge National Laboratory, U.S. Department of Energy, Oak Ridge, TN.

SEVEN

Recognizing Human Impacts

This exercise explores relationships between increasing human population size, demands for natural resources, and environmental degradation. Be sure to show all of your work when performing calculations.

SIMPLE MODEL FOR ENVIRONMENTAL DEGRADATION

A simple model of environmental degradation can be formulated as ED=PAT, where ED is environmental degradation, P is population size, A (affluence) is per capita resource use, and T (technology) is environmental degradation per unit resource use.[22] Environmental degradation, ED, increases or decreases with an increase or decrease in any of the three terms (P, A, and T). For a given level of P, unacceptable levels of environmental degradation can result by multiplying very large values of A by small values of T or by multiplying relatively small values of A by very large values of T. Excessive environmental degradation resulting from large values of P has been termed **people overpopulation**, whereas similarly excessive degradation because of large values of A has been termed **consumption overpopulation**. Let's examine some examples of how population size, resource use, and environmental degradation are interrelated.

1. In 1994, there were 630 million cars in the world and climbing fast (AAMA, 1996). In 2001, Americans registered 230 million cars. A substantial portion of the U.S. gross domestic product (GDP) is devoted to the automobile economy (motor vehicle manufacturing, gasoline, insurance, repair, road construction, parts, labor, and so forth). From 2002 to 2004, the population of the world was approximately 6.355 billion people and the U.S. population was approximately 293,000,000.

   a. What is the number of automobiles per capita for the world? (Assume 6.3×10^8 automobiles and 6.3×10^9 people.)

   b. What is the number of automobiles per capita for the U.S.? (Assume 2.3×10^8 automobiles and 293×10^6 people.)

   c. Comment on the difference between the per capita world and per capita American amount.

2. In 2002 (FHA, 2004), the U.S. consumed 137 *billion* gallons of gasoline. (Note that not all of this amount was for motor vehicles, but includes lawnmowers and generators,

---

[22]This model is a slight variation of the environmental impact model developed by Ehrlich, P.R. and Holdren, J.P. 1971. Impact of Population Growth. Science. 171:1212–1217.

although they represent a small percent of this amount.) The combustion of one gallon of gasoline produces 8.750 kg of CO_2 (ICBE, 2002).

a. What were the total annual CO_2 emissions for U.S. gasoline consumption in 2002?

b. What was the per capita generation of CO_2 emissions for the U.S. in 2002?

c. Assuming everyone in the world produced the same amount of CO_2 emissions as the U.S. did in 2002 from gasoline, what would the total worldwide CO_2 emissions be?

QUESTION OF QUALITY

Measuring environmental degradation ultimately comes down to evaluating how much the quality of life on our planet has declined. In the calculation you made above, you performed relatively straightforward calculations of CO_2 emissions, not direct measures of environmental degradation. Translating CO_2 emissions into environmental degradation is difficult. First, one would have to know the effect of the emissions in terms of increased global warming and all of its ramifications. Then all of the effects would need to be evaluated and assigned a cost, such as economic cost (cost in dollars) or health cost (loss in life or pain and suffering). Costs such as loss of a species or aesthetic degradation are difficult to quantify. A final evaluation of environmental degradation would ideally tell us how much the quality of life has declined.

3. In a paragraph, describe your idea for the best way to measure environmental degradation. What factors would you consider?

REDUCING PER CAPITA RESOURCE USE

One approach to reducing environmental degradation is to reduce per capita resource use through greater efficiency, conservation, greater durability of products, and recycling. The three Rs of resource use and Earth care (a.k.a. pollution prevention) can be thought of as reduce unnecessary consumption, reuse when possible, and recycle what you cannot reuse. (Recycling is important, but it is the least environmentally beneficial compared to source reduction and reuse.)

4. A typical toilet uses 18–26 liters (5–7 gallons) per flush (assume 22 liters per flush). Low-flush toilets use 6 liters (1.5 gallons) per flush. Assume that each of 10,000 university students flushes 5 times per day.

a. How many liters of water would be saved in one day if all toilets were low flush?

b. How many liters could be saved in a year?

5. During the 1980s, when a prolonged drought hit California, the following saying became common: "if it's yellow let it mellow, if it's brown flush it down." How do you feel about this?

6. The average student generates an estimated 290 kilograms (640 pounds) of solid waste (garbage) per year.

a. Multiply the average amount by the number of students to determine total student amount.

b. What about faculty and staff? Devise a way to determine their contribution for a total university amount.

c. What could be done to reduce this amount?

THE POPULATION FACTOR

Population growth (or decline) results from a balance of births, deaths, immigration, and emigration. Let's look at some basic calculations of population growth that don't require detailed information about age structure, survivorship, or fertility. We will use the exponential growth equation $dN/dt = rN$, where dN/dt is the rate of population growth, r is the intrinsic rate of growth expressed as number of new individuals per existing individual per time unit, and N is the number of individuals in the population. This equation can be solved, using calculus, to obtain the population size as it changes over time:

$N_t = N_0e^{rt}$, where N_t is the population size at time t, N_0 is the initial population size, e is base of the natural logarithm (a constant with a value of 2.718281828...), r is the intrinsic rate of growth, and t is time. For example, tiny town has a population of 1,100 and an r of 1.1%. What will the population be in 40 years? $1,100 \times (2.718281828^{((.011)(40))}) = 1,708$.

7. Given a population size in the U.S. of 293×10^6 individuals, an annual intrinsic rate of increase of 0.81% ($r = 0.0081$), what is the expected population size after 5 years?

8. The population size of Saudi Arabia is approximately 14×10^6 individuals, and the annual growth rate is approximately 3.96% (i.e., $r = 0.0396$ ind/ind/year). What is the expected population size after 17.5 years, assuming continued exponential growth?

INDIVIDUALS DO MATTER

It is easy to feel small and helpless—that environmental problems are so big that it does not matter what we as individuals do. At one extreme, this argument could be an excuse not to take responsibility for our actions and to justify doing whatever we want regardless of the environment impact. But we can make a difference if many of us act responsibly. As we have seen, small numbers multiplied by very large numbers can become large numbers. The slogan "think globally, act locally" is especially pertinent when working to solve environmental problems.

9. Globally, we use a lot of resources, but per capita resource use is different among different cultures.

  a. What would happen if each person on the Earth reduced his or her resource use by 25%?

  b. Is there a difference between a 25% reduction for an American compared to a 25% reduction for a Haitian?

10. Would the overall reduction of 25% mean anything in the greater scheme of the world if the current exponential growth of the human population were not controlled? Support your answer.

References

AAMA, 1996. American Automobile Manufacturers Association, Motor Vehicle Facts and Figures. AAMA, Washington, DC.

Federal Highway Administration (FHA). 2004. Highway Statistics. [Online] Available at http://www.fhwa.dot.gov/policy/ohim/hs02/mf21.htm (verified 9 June 2004).

International Carbon Bank Exchange (ICBE). 2002. Carbon Database. [Online] Available at http://www.icbe.com/CarbonDatabase/CO2volumecalculation.asp (verified 9 June 2004).

EIGHT

Ecological Footprints and Sustainability

This assignment allows the student to calculate his or her own "ecological footprint" and see how the footprint concept relates to sustainability.

INTRODUCTION

Like all species, humans need certain resources to survive. However, humans consume resources not only for survival, but also for comfort, luxury, and prestige. Whereas non-human species generally must obtain their resources from within their ecosystem, in contrast, humans have devised ways (transportation) to remove resources from other ecosystems to satisfy their wants and desires. However, societies are not equal in their ability to extract, transport, process, manufacture, and use resources. And, societies have different philosophies and cultural perspectives regarding their desire to utilize resources beyond basic needs. Thus, there is a question of equitable distribution of resources among human societies and between humans and other species.

In addition to resource extraction, an additional crucial ecosystem function is the assimilation of waste, sometime, known as sinks (e.g., air, water, and soil pollution; hazardous, solid, and radioactive waste; and waste heat). Again, humans have devised ways to discharge wastes into other ecosystems by building tall smoke stacks, dumping waste in flowing rivers and oceans, and shipping "recyclables" and wastes around the globe.

The area of productive land required to provide resources <u>and</u> assimilate waste to meet consumption needs is referred to as the **Ecological Footprint** (Wackernagel and Rees, 1996). This is different from **carrying capacity**, which is the maximum abundance of a population that can be sustained by a habitat or ecosystem without degrading the habitat or ecosystem. Because non-humans cannot extract resources from outside of their ecosystem, their population cannot exceed the carrying capacity, which is based on the availability and amount of an ecosystem's resources. Thus, a non-human's ecological footprint is limited by the size of the ecosystem. In contrast, humans also have a carrying capacity for their "ecosystem" (for example, a country). However, because humans can transfer resources from another country, their ecological footprint can exceed the carrying capacity.

Thus, the U.S.'s ecological footprint can exceed the carrying capacity of the U.S. (i.e., the U.S. can maintain more people than available resources) because resources are extracted from Mexico, Africa, Saudi Arabia, and so forth. Clearly, this means that for some countries, their ecological footprint must be smaller than the carrying capacity because the Earth is finite. Or, some populations must live near the subsistence level, whereas other can live in high comfort. A method to determine and compare this is to calculate

175

and compare the **per capita** amount of resource use (the amount available/consumed on a per person basis). Calculating the per capita is done by dividing the amount of available biological resources and waste assimilation needs by the population (resource ÷ population).

The ecological footprint is one measure of the sustainability of a society's current lifestyle. However, this is an **anthropocentric** view. If humans consume all the resources or take over all the biologically productive land, what about non-humans? And what about humans in less developed countries? This is an issue of **environmental equity**. As shown in Figure P8.1, the average per-person ecological footprint in the U.S. is nearly 10 times greater than the per-person ecological footprint of India.

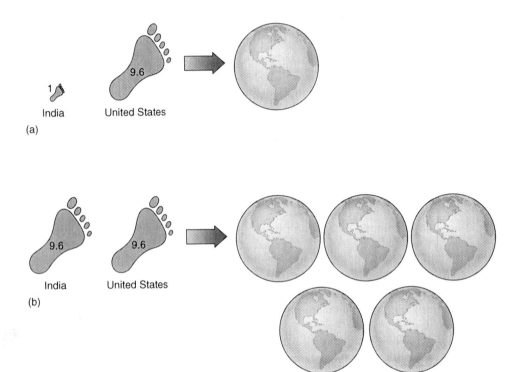

Figure P8.1—Comparison of Ecological Footprints Between India and the United States.
In developing nations such as India, about one hectare is needed to meet the resource requirements of an average person. In the U.S., the average individual ecological footprint is almost ten hectares. If everyone in the world consumed resources equal to the average American, we would need five Earths. Source: Raven, P.H. and L.R. Berg. 2004. Environment. 4ᵗʰ ed. John Wiley & Sons, NY (p. 189).

TASKS

In this lab, you will individually analyze your consumption and life activity patterns to calculate your ecological footprint using an Internet-based ecological footprint calculator. Please do the following tasks and answer the following questions:

1. Go to the U.S. Census Bureau's population calculator at http://www.census.gov/main/www/popclock.html

   a. What is the current world population? (Note the precise *local* time—for example, Eastern Standard Time—you determined this.)

   b. What is the current world person/hour growth rate?

   c. What is the current U.S. population? (Note the precise local time you determined this.)

   d. What is the current U.S. person/hour growth rate?

2. Go to http://www.earthday.net/footprint/index.asp or a site provided by your instructor. Answer all questions on the website truthfully. If you are not sure of the answer, make an educated guess.

   a. What is your ecological footprint in acres? in hectares?

   b. How does your footprint compare to the average American footprint? (Compare quantitatively as in percent difference.)

   c. If everyone in the world had the same footprint as you, how many Earths would be required to support the global population?

3. If you are using http://www.earthday.net/footprint/index.asp, click on the button "What About Other Species." The Bruntland Report states that 12% of the biosphere should be set aside for other organisms.

   a. By using 12%, what answer do you get for the number of Earths required?

   b. Do you think this amount (12%) should be higher or lower? Explain why.

   c. Assume that you believe that 25% of the Earth's resources should be set aside for non-humans. Now what is your answer?

4. Go outside to a field or parking lot indicated by your instructor. Using a measuring tape, calculate the area (in acres and hectares) of the field.

   a. How many acres are contained in the field? how many hectares?

   b. Based on your ecological footprint answer from Question 2.a., how many of these fields would be required to support your ecological footprint?

   c. What percentage of your campus would have to be devoted to supporting your life? What is your source?

5. The Earth's biosphere currently has an estimated 26.7 billion acres of biologically productive land.

   a. Now what is the world's current population? Use the U.S. Census Bureau's population calculator at http://www.census.gov/main/www/popclock.html

   b. At what precise local time did you determine this?

   c. How many people have been added since you first checked?

   d. Based on available biologically productive land and the current world population, what is the global per capita amount of biologically productive land? (Show your work.)

6. A key component to sustainability is population growth. Thus, it is important to estimate future population growth to assess its sustainability. One measure of future population is to estimate a population's doubling time, which is based on the current, annual percentage population growth rate. This is the so-called rule of 70, a rudimentary estimate. Divide 70 by the percentage of growth and you will obtain the doubling time. (For example, if growth is 2%, then 70/2 = 35 years.)

a. According to the U.S. Census Bureau, in 2003, the annual U.S. population increase was 1.01%. Assuming steady growth, in what year will the population of the U.S. double?

b. Approximately how many people will live in the U.S. on that date?

c. In 2002, the annual population growth rate for Mexico was 1.47%. Assuming steady growth, in what year will the population of Mexico double?

d. What will the population in Mexico be on that date?

e. What are the environmental implications of the doubling of population for Mexico and the U.S.?

7. What about equity? Is the U.S. ecological footprint fair and equitable compared to that of other countries? We will assume that the U.S. has 293.7 million acres of biologically productive land and that the average American needs 24 acres to maintain his/her current lifestyle. Now, go back to the U.S. Census Bureau's population calculator at http://www.census.gov/main/www/popclock.html.

a. What is the current U.S. population?

b. What is the precise local time you determined this?

c. How many people have been added since you first checked?

d. Is there enough biologically productive land in the U.S. for current consumption habits?

e. If not, where, specifically, do Americans obtain their resources?

f. What happens to the humans and non-humans living in those ecosystems?

g. What is the current world population?

h. What time did you determine this?

i. How many people have been added since you first checked?

j. Recalculate the per capita availability of global resources (i.e., resources divided by population).

k. Is the amount of global biologically productive land increasing, staying steady, or decreasing?

l. What are the ramifications of your answer based on the increasing population?

8. Assume steady growth, consumption rates, and size of the average ecological footprint for the U.S. Answer the following questions using some of the numbers you obtained in your answers. Explain the significance of population growth in relation to:

a. Ecological resources

b. Waste assimilation

c. Sustainability

d. Equity

9. Evaluate the ecological footprint calculator you used for this assignment. Do you think the footprint is accurate, why or why not? What could be done to improve the accuracy of the ecological footprint calculator?

Reference

Wackernagel, M. and W. Rees. 1996, Our Ecological Footprint: Reducing Human Impact on the Earth, New Society Publishers, British Columbia, Canada.

Oil Consumption and Future Availability

Oil is a non-renewable resource. In 1859, the first oil well—the Drake Oil Well—was drilled in Titusville, Pennsylvania, launching the petroleum revolution. On that day, we began to deplete the resource. Some argue that there is lots of oil sitting in the ground, but is it useable? If so, who should use it? Our most current National Energy Policy calls for more exploration and production of oil within the U.S. The rationale is that this will reduce our dependence on foreign sources of oil and increase our security. Questions we must ask ourselves include: "Is this merely political rhetoric?" and "Is this good environmental policy?" This assignment provides some background before we can answer these questions with confidence. Answer the following; be sure to give the source for any data or facts you cite.

AVAILABLE OIL

There is widespread lack of understanding of the significant difference between the terms "resources" and "reserves."

The total *resource* base of oil is the entire volume formed and trapped in-place within the Earth before any production. The largest portion of this total resource base is non-recoverable by current or foreseeable technology. Most of the non-recoverable volume occurs at very low concentrations throughout the Earth's crust and cannot be extracted short of mining the rock or the application of some other approach that would consume <u>more</u> energy than it produced.

Reserves, a subset of the total resource base that is of societal and economic interest, are technically recoverable portions of the total resource base (See Table P9.1) Therefore, reserves are not a fixed amount and are estimated; they can increase or decrease based on geologic data and technology.

| TABLE P9.1 | Technically Recoverable Petroleum Resources in U.S. as of 2002 (EIA, 2003) |

Location	Ownership	Billion Barrels
Alaska Onshore + State Offshore	Federal	3.75
Alaska Onshore + State Offshore	Other	4.68
Alaska Federal Offshore	Federal	24.90
Lower 48 States Onshore + State Offshore	Federal	3.79
Lower 48 States Onshore + State Offshore	Other	17.83
Lower 48 States Federal Offshore	Federal	50.10

1. How many billion barrels (42 gallons each) of technically recoverable oil resources are there in the U.S.?

2. With a 95% confidence level, how much technically recoverable oil resources exist in Alaska's Arctic National Wildlife Refuge (ANWR)[23]?

3. What is the mean level of technically recoverable oil resources in ANWR?

4. Search the Internet and determine the mean number of miles traveled per vehicle in the USA last year. What is the average mpg for a passenger car in the past year? What is it for a truck? Look at the changes in mpg for cars and truck over the past 20 years and determine overall percentage of change. Hint: http://www.eia.doe.gov.

5. Using the Internet for research, create a table comparing the gallons of gasoline consumed in the U.S. for passenger cars every year since 1970 to the most current reporting year. Comment on the changes.

6. How long would ANWR supply our gasoline needs? (Hint: use Table P9.2 and your answers to Questions 2 through 5.) Support your answer with calculations and/or narrative.

| TABLE P9.2 | Petroleum Products Produced from a 42-Gallon Barrel of Crude Oil (SCPMA, 2004) |

Product	Gallons per Barrel
Gasoline	19.5
Fuel oil (includes home heating oil and diesel fuel)	9.2
Jet fuel	4.1
Residual fuel oil (heavy oils used as fuels in industry, marine transportation and for electric power generation)	2.3
Miscellaneous (liquefied refinery gasses, still gas, coke, asphalt and road oil, petrochemical feedstocks, lubricants and kerosene)	6.9

[23]One source of information is the U.S. Department of Energy publication (2000), Potential Oil Production from the Coastal Plain of the Arctic National Wildlife Refuge: Updated Assessment available at http://www.eia.doe.gov/pub/oil_gas/petroleum/analysis_publications/arctic_national_wildlife_refuge/html/anwr101.html

7. Using Table P9.2 and your answers to Question 1, assuming all the recoverable oil is converted to U.S. supply, how long can we supply our gasoline needs using domestic sources?

8. What are the likely environmental and political implications of our current path?

9. Provide some suggestions on how we should supply our increasing need for gasoline.

10. What are some realistic alternatives to reduce consumption by the private sector and government?

References

U.S. Energy Information Administration (EIA). 2003. Annual Energy Review, 2002. Department of Energy, Washington, DC,

South Carolina Petroleum Marketers Association (SCPMA). 2004. [Online] Available at http://www.scpma.com/faqs.htm#barrel (verified on 9 June 2004).

TEN

Water Quality, Chemicals, and Consumer Choice

This assignment is designed to introduce students to thinking about emerging water quality issues and the role of the precautionary principle in the decisions we make about what products to market or use.

INTRODUCTION

A primary goal of environmental science is to identify and examine emerging environmental concerns before they become critical ecological or human health problems. In 1996, the book *Our Stolen Future* (Colborn et al.), raised concerns about the prevalence of certain synthetic chemicals in the environment that can interfere with hormonal messages involved in the control of human and nonhuman growth and development, especially in the fetus. (These are referred to as endocrine disruptors or hormone mimickers.) This book helped focus attention on these and other ubiquitous pollutants in the environment, which had not been studied to the same degree as so-called conventional water pollutants because of their expected low levels. These "overlooked" environmental contaminants include reproductive hormones, steroids, antibiotics, pharmaceuticals, personal care products, detergents, disinfectants, fragrances, insect repellants, and fire retardants. These compounds and their metabolites have been detected in European waters and, more recently, in U.S. streams. (**Metabolites** are byproducts produced from metabolism, the organic processes necessary to sustain life.)

To date, little attention has been paid regarding the potential impact of pharmaceuticals and personal care products on ecological or human health. This concern is exacerbated by a growing population, increased consumption of these products, and increasing worldwide demand for safe freshwater. These compounds enter the environment directly through discharges from public sewage treatment plants and indirectly through wet weather runoff from animal feed lots and excreta from medicated domestic pets (Daughton, 2003).

According to Kolpin et al. (2002), "Surprisingly, little is known about the extent of environmental occurrence, transport, and ultimate fate of many synthetic organic chemicals after their intended use, particularly hormonally active chemicals, personal care products, and pharmaceuticals that are designed to stimulate a physiological response in humans, plants, and animals." A contributing factor to this lack of data is the lack of analytical methods capable of detecting these compounds at the low concentrations expected in the environment. The environmental presence of these compounds raises many concerns (e.g., abnormal physiological processes, reproductive impairment, cancer, and toxicity), but perhaps the biggest concern is that the compounds can contribute to the development of antibiotic-resistant bacteria. If super bacteria evolved through resistance to antibiotics, how would we defend ourselves from harmful bacteria?

Kolpin et al. (2002) conducted the first nationwide reconnaissance of the occurrence of pharmaceuticals, hormones, and other organic wastewater contaminants in water resources. Using a list of 95 contaminants, they identified 139 streams susceptible to contamination across 30 states during 1999 and 2000. Of the assessed streams, 80% had one or more of these contaminants. Some of the most frequently detected compounds were:

- N,N-diethyltoluamide (also known as DEET, used as a topical insect repellant)
- Caffeine
- Tri(2-chloroethyl)phosphate (fire retardant used in plastics)
- Triclosan (antimicrobial disinfectant)
- 4-nonylphenol (an ingredient in certain detergents)

For this assignment, we will focus on two compounds of interest: 4-nonylphenol and triclosan.

4-Nonylphenol

Although nonylphenol compounds used in detergents have been phased out in Europe, they are still widely used in the United States. Of special concern is 4-nonylphenol, which is one of the wide varieties of environmental chemicals reported to have estrogenic effects, which are thought to mimic the hormone estrogen and potentially affect human and non-human reproduction and development.

Triclosan

Triclosan is used in "antibacterial" and "antifungal" detergents, dishwashing liquid, laundry detergent, deodorants, cosmetics, lotions, creams, toothpastes, and mouthwashes. There are concerns that the widespread use of Triclosan may promote antibiotic-resistant bacteria. Triclosan is a chlorophenol, a class of chemicals classified as a suspected animal carcinogen. Clorophenols are also categorized as a persistent organic pollutant (POP), meaning that it can persist in the environment and bioaccumulate up the food chain.

TASKS

1. Go to a local pharmacy or grocery store. Consider seven types of consumer products presented in Table P10.1. Look at several brands and try to locate one that contains triclosan and/or nonylphenols.

TABLE P10.1	Common Consumer Products		
Product	**Brands**	**Contains Triclosan (yes or no)**	**Contains Nonylphenol (yes or no)**
Hand Soap			
Dishwashing Detergent			
Laundry Detergent			
Skin Lotion			
Toothpaste			
Mouthwash			
Deodorant			

    a. Using the product's label, how much information is provided on the ingredients? That is, can you tell what they are or their potential environmental or health risk? Is the information sufficient to make an informed decision (regarding environmental and personal safety) when purchasing the product?

    b. Based on your above table, how common are these two compounds in consumer products (e.g., percent)?

    c. On average, how much of each of the seven products in the above table do you use per month?

    d. Multiply these amounts by 12 and then by 293 million for a rough approximation of the total U.S. consumption of these products per year.

    e. Explain how valid this technique is for a rough estimate.

    f. For <u>each </u>of these seven products, describe how they enter the surface water (step-by-step) during and after use.

    g. Conduct an Internet or database search for news items containing the following terms: hormone mimicker, endocrine disruptors, triclosan, and nonylphenol. Report what you found.

2. Currently, there is little evidence to indicate that these two compounds present a direct, significant health or ecological risk. This, however, is because of a lack of data and the low levels observed in the environment. Although additional research may indicate that these compounds do not present an appreciable health threat, some advocate the adoption of the precautionary principle when dealing with widespread environmental contaminants. The **precautionary principle**, initiated in Europe in the 1970s and articulated in the Rio Declaration of the 1992 United Nations Conference on Environment and Development (Agenda 21), means to take prudent action when there is sufficient scientific evidence (but not necessarily absolute proof) that inaction could lead to harm and where action can be justified on reasonable judgments of cost-effectiveness (Foster et al. 2000).

    a. What specific type of scientific evidence do you believe is necessary to support a ban?

   b. List and explain the various considerations policymakers would need to make in banning these two compounds.

   c. List at least three other specific emerging environmental issues/problems in which we have scientific evidence, but have not yet acted to any appreciable degree.

   d. Why have we failed to act in these cases?

3. Assume there will be no ban on hormone mimickers in the near future.

   a. What consumer choices can you make that will likely reduce the potential impact on the environment?

   b. What other available information on a consumer product can help make environmentally informed choices?

References

Colborn, T., Dumanoski, D., and J.P. Myers. 1996. Our Stolen Future: Are We Threatening Our Fertility, Intelligence, and Survival? Penguin Books, New York.

Daughton, C.G. 2003. Environmental Stewardship of Pharmaceuticals: The Green Pharmacy. In Proceedings of the 3rd International Conference on Pharmaceuticals and Endocrine Disrupting Chemicals in Water, National Ground Water Association, 19–21 March 2003, Minneapolis, MN.

Foster, K.R., Vecchia, P. and M.H. Repacholi. 2000. Science and the Precautionary Principle. Science 288:979–981.

Kolpin, D.W., Furlong, E.T., Meyer, M.T., Thurman, E.M., Zaugg, S.D., Barber, L.B., and H.T. Buxton. 2002. Pharmaceuticals, Hormones, and Other Organic Wastewater Contaminants in U.S. Streams, 1999–2000: A National Reconnaissance. Environmental Science and Technology 36:1202–1211.

ELEVEN

Local Environmental Risk

This assignment is designed to introduce you to understanding the possible risks in your own community.

Go to http://www.scorecard.org/ and type in your zip code and answer the following questions:

1. What is your zip code?

2. What county and state is this?

3. In your own words, what is cancer?

4. Define human health risk.

5. What percentage of air cancer risk is from mobile sources? What is a mobile source?

6. What percentage of air cancer risk is from area sources? What is an area source?

7. What percentage of air cancer risk is from point sources? What is a point source?

8. Based on the Pollutant Standards Index, how clean is your air?

9. In the section on Toxic Chemical Releases from Manufacturing Facilities, how many facilities are listed?

10. Select one of these facilities. What is the address? What do they produce? Do you know someone who works there?

11. How many Superfund sites are there?

12. In your own words, what is a Superfund site?

13. What is the Clean Water Act status of your county? That is, to what degree do water bodies in your community meet Clean Water Act standards?

14. Many suggest that there is inequity in exposure to environmental risks. That is, different segments of the population are exposed to greater environmental risks than others. Who are these people who may have increased exposure and why?

15. What are some criticisms of this website? That is, do you believe everything it says? Why or why not?

16. What other factors not listed on this website would have a significant impact on your risk of injury. Do NOT limit these to pollution, but <u>list</u> all other environmental and non-environmental factors potentially impacting your health.

17. Check another zip code where a friend or relatives live. What is their risk?

18. What are the major sources of your own environmental risk? How much of these sources are within your control?

19. Find and provide the Internet site or address of another organization that provides current environmental information related to health, risk, or environmental quality. What is the organization, and what does it address?

```
PROBLEM SET
```

```
TWELVE
```

Society and Waste

The average American household produces a substantial amount of waste (Figure P12.1). However, this is not the only form of waste; there is also a far larger amount of industrial waste produced. This assignment is designed to introduce you to understanding the amount and types of waste generated in the U.S. and your own community.

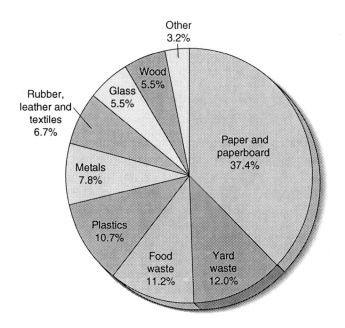

Figure P12.1—Composition of Municipal Solid Waste.
Source: Raven, P.H. and L.R. Berg. 2004. Environment 4th ed. John Wiley & Sons, NY (p. 545).

A. In 2001, U.S. residents, businesses, and institutions produced 229×10^6 tons of Municipal Solid Waste (MSW) (US EPA, 2003a).

    1. Assuming a population of 293×10^6, what was the per capita generation rate: (a) per day, (b) per week, and (c) per month in kilograms?

B. In 2001, the percentage composition of various categories of MSW by weight was (US EPA, 2003a):

Paper: 35.7%
Yard Waste: 12.2%
Food Waste: 11.4%
Plastics: 11.1%
Metals: 7.9%
Rubber, Leather, and Textiles: 7.1%
Glass: 5.5%
Wood: 5.7%
Other: 3.4%

  2. How many tons of each category of MSW was produced in 2001?

  3. Construct a table with the following headings:

Category	Amount	Examples

  4. In the above table, for each category of waste, provide five examples in the associated examples column.

C. Pollution prevention is a hierarchical approach to waste management based on the three Rs: reduce, reuse, and recycle.

Reduce, or source reduction, means to reduce or eliminate the creation of pollutants (i.e., reduce the volume and/or toxicity) through equipment or technology modifications, process or procedure modifications, reformulation or redesign of products, substitution of raw materials (with less hazardous materials), and improvements in housekeeping, maintenance, training, or inventory control.

Reuse means the reuse of a material without processing the material. (This does not mean that the material has to be reused for its original intent; it can be reused for other purposes provided it does not undergo significant processing. For example, constructing an artificial reef for fish habitat with old tires does not require significant processing (e.g., remelting) of the tires.

Recycle is the least preferred action under the pollution prevention hierarchy because it requires significant processing, which requires energy and produces pollution, although less than production from raw materials (e.g., collecting used paper to be recycled requires major processing and energy).

  5. In a table, for <u>each</u> waste category, identify a technique to reduce, reuse, and recycle. For example:

Paper	**Reduce = Double-sided copying**
	Reuse = Use used paper for scrap paper
	Recycle = Collect used paper to recycle into new paper

D. A new approach to reducing the generation of MSW is called unit-based pricing, or "pay-as-you-throw." Traditionally, residents pay for waste collection through property taxes or a fixed fee regardless of the amount generated. Pay-as-you-throw (PAYT) treats trash services just like electricity, gas, and other utilities. Households pay a variable rate depending on the amount of service they use; I_M in this case, the amount (volume or weight) of waste they produce. Most communities with PAYT charge residents a fee for each bag or container of waste generated. In a small number of communities, residents are billed based on the weight of their trash. In 1990, 148 communities serving a population of 1,009,000 had adopted a PAYT system. In 1999, the number was 4,032 communities serving a population of 35,049,000 (Miranda, 1999).

   6. Does your community have a PAYT program? If not, where is the closest community? What is the per can or per bag charge?

   7. What was the percentage increase for number of programs and population served in the U.S.?

   8. Explain why the number of PAYT programs has likely increased (i.e., discuss likely contributing social, economic, and technological factors).

E. In 2001, the U.S. generated 40.8 million tons of waste legally defined as hazardous (US EPA, 2003b). (Hazardous waste is defined as corrosive, reactive, toxic, or ignitable.)

   9. Assuming a population of 293×10^6, what was the per capita generation rate (a) per day, (b) per week, and (c) per month (in kilograms)?

   10. How does this compare (numerically) to MSW?

F. Each year, industrial facilities generate approximately 7.6 billion tons of nonhazardous industrial waste, most of which is managed in landfills. Generated by a broad spectrum of U.S. industries, industrial waste is process waste associated with manufacturing and is not classified (on a legal basis) as either municipal waste or hazardous waste by federal or state law.

   11. Assuming a population of 293×10^6, which was the per capita generation rate (a) per day, (b) per week, and (c) per month (in kilograms).

   12. What is the total annual per capita generation rate in tons for all waste (MSW, hazardous, and, nonhazardous industrial waste)?

   13. Explain the connection between MSW, hazardous waste, and industrial waste.

References

Miranda, M.L. 1999. Unit-Based Pricing in the United States: A Tally of Communities, Duke University. [Online] Available at http://www.epa.gov/epaoswer/non-hw/payt/research. htm (verified 7 June 2004).

U.S. Environmental Protection Agency (US EPA). 2003a. Municipal Solid Waste in the United States: 2001 Facts and Figures. U.S. Environmental Protection Agency, Washington, DC.

U.S. Environmental Protection Agency (US EPA). 2003b. National Analysis, The National Biennial RCRA Hazardous Waste Report (based on 2001 Data). U.S. Environmental Protection Agency, Washington, DC.

Introduction to Environmental Modeling

INTRODUCTION

Modeling is a tool to simulate or recreate reality. An **environment model** is a tool specifically designed to simulate or recreate the environment or, more specifically, an environmental system. It is often easier and less expensive to work with models compared to the actual system. However, models are valuable only if they are properly constructed and are fed good data; the popular saying "garbage in garbage out" applies to modeling.

Models are generally of two types: static and dynamic. **Static models** are used to understand the behavior of a system at rest. Economists use static models extensively. **Dynamic models** allow us to examine a system over time and are used by environmental scientists to examine changes to an ecosystem. Models have three basic components: the underlying science, a mathematical representation of the science, and a solution of the mathematics.

This Problem Set provides you with the opportunity to explain the basic concepts of modeling and use a model to make determinations about an environmental system. It should help you be able to describe several major challenges facing environmental regulators.

TASKS

Stock and flow modeling is the most basic form of dynamic environmental modeling. As shown in Figure P13.1, an example of a stock and flow is a human population. You have births and immigrants flowing in (inflow), a population (stock), and deaths and emigrants flowing out (outflow).

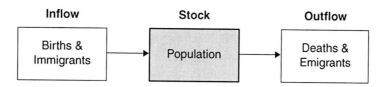

Figure P13.1—Simple Stock and Flow Model.

Based on the information below, you will be modeling the level of a particular contaminant in a pond and answering a series of questions based on use of the model.

Scenario:

- The Copper Brothers Manufacturing Company is located on the western shore of Valley Pond as shown in Figure P13.2.

Figure P13.2—Valley Pond Study Area.

- The pond's volume is 4×10^7 m³ of water.

- The average flow-through rate is 8×10^4 m³/day. That is: (1) the inflow from Little Valley Stream, (2) the water being discharged from the company into the pond, and (3) feeder springs collectively equal the volume of the outflow in Big Valley Stream (i.e., inflows = outflow) at 8×10^4 m³/day.

- The company produces decorative copper art by chemically etching the copper with strong sulfuric acid.

- The plant has a National Pollutant Discharge Elimination System (NPDES) permit issued by the state under the Clean Water Act. The NPDES permit allows the plant to discharge 0.16 tonnes (metric tons) of copper sulfate per day. The plant has an exemplary record of permit compliance. Currently, 25 people are employed at Copper Brothers. The company is the primary employment base for the town of Valley View.

- A family purchased a small camp on the eastern shore of Valley Pond. Over the summer, on numerous occasions, they observed dead fish in their tiny cove near Big Valley Stream. They contacted the State Water Quality Division to file a formal complaint against Copper Brothers.

We need to calculate the steady-state level of copper sulfate in Valley Pond to determine if there is too much in the pond. (That is, how much copper sulfate is in the pond given inflows, outflows, and copper sulfate discharges?) Although the plant is in compliance with their permit, the level of copper sulfate may be too high, biologically, for certain fish species because in the pond, it may be increasing, but it also may be decreasing. This is a function of the accumulation of copper over time.

The rate at which copper sulfate is added to the lake is known (0.16 tonnes—metric tons per day). So, to calculate the steady-state stock of the pollutant, we need to know its residence time in the pond.

We will assume that the pollutant is uniformly mixed in the pond and is highly water-soluble. (As with all models, certain assumptions must be made.) Thus, the residence time of the pollutant is equal to the residence time of the pond water. We can calculate the residence time of the water as:

Residence time: $T_w = M_w/F_w$

T_w = residence time of water in the pond

M_w = stock of water: the pond volume

F_w = average, daily flow through rate of the water

1. What is the residence time of the pond water?

The steady-state stock of copper sulfate can be calculated based on the following formula:

Steady stock: $S_{cs} = F_{cs}*T_{cs}$

S_{cs} = steady-state stock of copper sulfate

F_{cs} = daily discharge amount of copper sulfate

T_{cs} = residence time of the copper sulfate (see T_w)

2. What is the steady-state stock (load) of copper sulfate in Valley Pond?

3. What is copper sulfate? And what are its likely effects on Valley Pond?

The state's environmental standard for copper sulfate in aquatic systems is 1.98 parts per million. That is, 1.98 parts of copper sulfate per one million parts per water is allowed. We need to calculate the concentration of the copper sulfate in the water. This rather simple calculation requires us to divide the steady-state stock of copper sulfate by the total volume of Valley Pond (then multiply the number by 1×10^6):

$$C_{cs} = S_{cs}/M_w$$

4. What is the concentration of copper sulfate in Valley Pond (expressed in ppm)?

5. Based on the concentration, as a state environmental scientist, what would you communicate to the Chief of the Water Quality Division?

The Copper Brothers Manufacturing Company asserts that if it is prevented from discharging copper sulfate into Valley Pond, it will have to close, which will put all 25 people out of work. The company hires an engineering consultant to investigate manufacturing alternatives. The consultant concludes that Copper Brothers cannot reduce the quantity of copper sulfate it discharges without adversely affecting product quality. The company hires a consulting environmental engineer to present an ecological alternative, which the company proposes to the state. The company proposes to construct a pipeline that will connect Mountain Pond to Valley Pond, which, they argue, would dilute the copper sulfate below any adverse level. Essentially, the project would drain Mountain Pond and increase the volume of Valley Pond. Thus, as designed, the pipeline will increase the volume (but not the flow) of Valley Pond by 1 million m^3 (1×10^6 m^3). Using this assumption, answer the following:

6. What is the revised residence time of the copper sulfate?

7. What is the revised, predicted steady-state stock (load) of copper sulfate in Valley Pond?

8. What is the recalculated concentration (ppm) of copper sulfate? Will this proposal meet the state's water quality standard?

9. What are some likely environmental effects (i.e., unintended consequences) of increasing the water volume of Valley Pond by using Mountain Pond?

10. Is this proposal likely to appease the owners of the camp? Why or why not?

You ask a colleague to review the consultant's report. Your colleague notes that there is a glaring error in the model's assumptions: the consultant did not take into account evaporation! Evaporation will have a significant impact on copper sulfate concentration because evaporating water contains no copper sulfate. Thus, the steady-state concentration of copper sulfate would expect to be significantly higher because one of the possible exit pathways (evaporation) is no longer available. (That is, the flow-through rate and thus the residence time are wrong.) Therefore, the residence time of the copper sulfate is no longer equal to the residence time of the water, but rather the residence time associated only with the outflow of Big Valley Stream. Based on some rough calculations, the total rate at which water exits Valley Pond is: 33% by evaporation and 66% through Big Valley Stream.

11. What is the revised residence time of the copper sulfate <u>without</u> considering the proposed pipeline? (Remember, the original flow-through rate is reduced by 33%.)

12. What is the revised, predicted steady-state stock (load) of copper sulfate in Valley Pond?

13. What is the recalculated concentration (ppm) of copper sulfate? How does this amount relate to the state's water quality standard?

The state is concerned that this modeling approach be reevaluated for its ability for accuracy. The state regulators ask for general comments regarding the model.

14. What are some additional inflows and outflows that should be considered?

15. What are some factors that could affect the behavior of water pollutants that could have a significant effect on biota, which were not incorporated into the model (e.g., review the assumptions)?

16. What are some environmentally less damaging alternatives that the Copper Brothers could employ?

17. As a state environmental regulator, what are some major challenges that you would face in determining levels of pollution in water bodies using models?

FOURTEEN

Ecological Identity

The purpose of this exercise is for you to think about your own sense of ecological identity. An *ecological identity* is the sum of the various ways you see yourself in relationship to the Earth—how it is reflected in your personality, actions, values, and behavior. It refers to your connection to the Earth, your understanding of ecosystems, and your past experiences of nature, which collectively contribute to your sense of self (Thomashow, 1995). Ecological identity is a process of constructing meaning and an ecological worldview that can promote personal change.

Answer both of the following:

1. Look at the Environmental Awareness activity used in the lab manual. Create a set of 10 questions to promote environmental awareness about a place (e.g., your town, your tribal lands, central New York, Okefenokee Swamp, Missouri) or a resource (e.g., lakes). Select questions that could contribute to ecological identity—yours or for someone else developing their own. Provide the answers and their sources.

2. How have your knowledge and sense of the environment contributed to your own ecological identity? Write a two-page essay about your ecological identity, including its role in your future, by citing actions, activities, events, and direct experiences involving this course and how your ecological identity will influence your future.

Reference

Thomashow, M. 1995. Ecological Identity: Becoming a Reflective Environmentalist. The MIT Press, Cambridge, MA.

Review and Reflection

This Problem Set reviews the main points of the previous Problem Sets, examines how you as an individual matter, and prompts you to look at the big picture.

WHAT DOES IT ALL MEAN?

In the first lab we began to cultivate our environmental awareness and explored where and how we fit. The first two Problem Sets focused on the scientific method and its application in environmental science. From there, Problem Sets focused on how to quantify environmental problems and the impact of humans on environmental quality, in particular population size, resource use, pollution, and other environmental degradation. Throughout the course, the Problem Sets have encouraged you to examine your role in the environment. What are the consequences of choices you make? or the way you choose to live your life? Look back over the Problem Sets and think about how they all fit together.

SHORT ESSAYS

1. Summarize each of the Problem Sets and explain how they fit into the basic model of environmental degradation (Environmental degradation = P $\times$ A $\times$ T, where ED is environmental degradation, P is population size, A (affluence) is per capita resource use, and T (technology) is environmental degradation per unit resource use[24]).

2. Describe how the Problem Sets and laboratory exercises have affected your understanding of environmental science from personal and academic perspectives.

[24]This model is a slight variation of the environmental impact model developed by Ehrlich, P.R. and Holdren, J.P. 1971. Impact of Population Growth. Science. 171:1212–1217.

PART FOUR

Appendices

Glossary

Abiotic factors: nonliving or physical factors (temperature, light) in the environment.

Air exchange rate: the rate at which outdoor air replaces indoor air in a building or room.

Air pollution: contamination of the air through the release of harmful substances.

Ambient air quality: the overall condition of the outside air (compare to indoor air).

Anthropocentric: a human-centered worldview.

Anthropogenic: human-made pollutants.

Ash fill: landfills devoted to disposal of municipal solid waste incinerator ash.

Berm: raised area, usually vegetated, often a buffer for a variety of uses such as aesthetics, noise reduction, erosion control, wind screening, and flood control.

Biodiversity: the different species and life-sustaining processes that can best survive the variety of conditions found on Earth.

Biotic factors: living factors in the environment.

British Thermal Units: BTU, the amount of energy required to raise the temperature of 1 lb of water 1° Fahrenheit when the water is near 39.2° Fahrenheit.

Caloric content: the amount of food calories in a given food item.

Carbon dioxide measurements: CO_2 meter used to measure the quantity of carbon dioxide in the air. Sustained levels above 1000 ppm can indicate inadequate ventilation indoors.

Carbon dioxide production: the amount of carbon dioxide produced; the average gallon of gas produces 9 kilograms of CO_2.

Carrying capacity: the maximum abundance of a population that can be maintained by a habitat or ecosystem without degrading the habitat or ecosystem.

Concentration-response curve: the method used to document the effects of various concentrations of a chemical substance on a group of organisms.

Contamination: the tainting of an item (i.e., soil) through human activities, including the intentional and unintentional discharge of hazardous materials and waste.

DDT: Dichlorodiphenyltrichloroethane, an insecticide used mostly in the mid-20th century, it is credited with saving hundreds of thousands of human lives around the world, but is now severely restricted in the U.S. due to its detrimental effects on the environment.

Degree day: a measure of the difference between the mean daily temperature and a given standard (often 65° F), and used to calculate heating and cooling requirements.

Ecological diversity: variety of habitats (forests, grasslands, streams).

Ecological footprint: the area of productive land required to provide resources and assimilate waste products to meet our consumption needs.

Energy investment in food production: the amount of energy that goes into producing, processing, packaging, and shipping food.

Environmental audit: inspection of a designated area to determine its overall environmental health.

Environmental equity: the fair distribution of land and resources between humans and other organisms.

Environmental model: a tool specifically designed to simulate or recreate the environment or, more specifically, an environmental system.

Environmental risk: the potential to cause harm to human health and/or the environment.

Environmental toxicology: the study and detection of environmental poisons and their effects on humans and the environment.

Exhaust emissions: the fumes released from vehicles including sulfur oxides, nitrogen oxides, volatile organic compounds, particulate matter, carbon monoxide, and lead.

Experimental design: creation of or description of an experiment through which a hypothesis can be tested and which adheres to the scientific method.

Exposure: the potential for a person to come into contact with a contaminant.

Field screening techniques: the techniques used in determining the source of an environmental problem, including specific conductance, dissolved oxygen, turbidity, ammonia, and pH.

Global environment: the environment of the entire world.

Global sustainability: efficient use of resources to sustain the current population and allow for the use of resources for future generations.

Habitat: local environment in which an organism, population, or species lives.

Heat loss: loss of heat due to inefficient or insufficient insulation.

Hypothesis: a tentative statement that proposes a possible explanation to some phenomenon or event.

Hypothesis testing: testing a hypothesis through a designed experiment that follows the scientific method.

Indoor air quality: the nature of air that affects the health and well-being of occupants.

Infiltration: the process by which outdoor air flows into a building or room through openings, joints, and cracks in walls, floors, and ceilings, and around windows and doors.

Kilowatt-hours: kWh, unit of energy equivalent to one kilowatt of power expended in one hour; not a standard unit, but is commonly used in electrical applications.

Laboratory analysis: the analysis of data through observation and experimentation in a laboratory.

Landfill: a plot of land used for the long-term storage solid waste.

Lethal concentration: concentration of a chemical that causes death to organisms; most commonly measured at LC_{50}, the concentration that kills half of the treated organisms. The lower the LC_{50} value, the greater the toxicity.

Lethal dose: as opposed to concentration, dose is given directly to the organism, such as by injection or orally.

Life cycle assessment: a process to identify and evaluate the potential environmental effects of a product over its lifetime.

Mean: the arithmetical mean—the average. To calculate mean, sum up the individual numbers and divide by the total count of how many numbers you summed.

Mechanical ventilation: uses devices, such as outdoor-vented fans, to intermittently remove air from a room, such as from bathrooms, to air handling systems that use fans and duct work to continuously remove indoor air and conversely distribute filtered and conditioned outdoor air to strategic points throughout a building or house.

Modeling: a tool to simulate or recreate functional reality to an approximate but acceptable degree.

Municipal solid waste: solid waste that includes trash, garbage, rubbish, and refuse.

Natural ventilation: air moving through opened windows, doors, and passive vents.

Non-renewable energy: energy sources that cannot be replaced once they have been used.

Null Hypothesis: a hypothesis that states the variables do not have an effect on the outcome of the experiment.

Nutritional value: the amount of nutrients (vitamins, minerals) in a given food item.

Parent material: the material from which a soil is formed.

Passive solar energy: capturing sunlight directly and converting it to low-temperature heat used domestically for heating air and water; does not use any mechanical devices.

Pedosphere: the thin layer of soil on the Earth.

Peer review: a process used for checking and verifying the work performed by one's equals—peers—to ensure it meets specific academic and scientific criteria.

Per capita energy consumption: total amount of energy consumed per person.

Personable accountability: recognizing and taking personal responsibility for the resources that an individual consumes.

Phase I Site Assessment: the first phase of an environmental site inspection; includes basic observations of the environmental health of a designated area.

Phase II Site Assessment: seeks to verify the environmental observations identified in a Phase I Site Assessment and delineate the problems through sampling and analysis.

Phase III Site Assessment: employed to remediate (clean up) delineated contamination.

Phytotoxicity: poisonous to plants.

Pollution prevention: preventing pollution through reducing waste generated, reusing otherwise disposable products, and recycling materials.

Pollution Prevention Act of 1990: a congressional act which established a national policy that includes: preventing or reducing pollution at the source wherever possible, reusing and recycling pollution in an environmentally safe manner whenever feasible, treating pollution which cannot be prevented or recycled, and allowing disposals into the environment only as a last resort and in an environmentally safe manner.

Population: a group of individuals of the same species living in the same area.

Population growth: the growth of a population based upon total births, deaths, immigration, and emigration.

Primary treatment: the initial treatment applied, particularly regarding sewage treatment, which utilizes bacteria to degrade waste.

P-value: probability that a test statistic in a hypothesis test captures the actual difference between groups (treatment and control) during an experiment.

Rachel Carson: writer of *Silent Spring* and other notable books. Carson was an environmentalist who opposed the misuse of pesticides, particularly DDT.

Range of tolerance: the varying range of environmental conditions a species can tolerate; individuals within a species may also have slightly different ranges of tolerance.

Recycle: reducing pollution by reprocessing materials to a new usable form.

Reduce: the reduction of pollution by using fewer materials.

Research techniques: procedures used in gathering information.

Reuse: reducing pollution by reprocessing, recycling through a system, or otherwise reemploying materials such as containers or water in the cooling systems of facilities.

Risk perception: an individual's assessment, based on feeling or judgment, for the potential harm of a substance, event, or setting.

Scientific method: the process of experimentation which includes making an assumption about an observation or phenomenon, creating a hypothesis, testing the hypothesis, and accepting or rejecting the hypothesis based on results of an experiment, and finally revising the hypothesis.

Secondary treatment: the second treatment applied to wastewater (sewage), to insure the degradation of waste before it is expelled into the ocean.

Sequential comparison index: a quick method to ascertain a rough picture of species diversity without having to identify the specific species.

Sick building syndrome: a set of symptoms that affect some occupants during time spent in a building and diminish or go away during periods away from the building, but cannot be traced to specific pollutants or sources within the building.

Soil disturbance: a disturbance in the soil such as burrowing or filling.

Soil horizon: the layers that make up a soil profile, each horizon identified by changes in color or texture.

Soil profile: the way a soil looks at a cross-sectional view.

Soil texture triangle: the three variables (sand, silt, and clay) used in determining soil type based on texture.

Species evenness: relative proportion of each species.

Species richness: number of different species.

Steady-state model: type of model that the stock (such as a pond) does not change, or is balanced by equal inflow and outflow.

Superfund: officially known as the Comprehensive Environmental Response, Compensation, and Liability Act; establishes strict liability for property owners where hazardous substances present a threat to human health or the environment, regardless of when the hazardous substances were placed there.

Survivorship curve: a graphical representation of the likelihood that an individual will survive from birth to a particular age.

Testable: a test of how two variables might be related.

Toxicity: the degree at which a chemical becomes toxic.

Toxicology: study and detection of poisons and their effects.

t-test: statistical test for comparing the means values from two samples. It is used to show how confident one can be that the two mean values differ.

Waste-to-energy facility: a facility that incinerates waste and uses the heat energy produced to power generators, creating electric energy usable by consumers.

The Metric System

The metric system originated in the 1790s as an alternative to the peculiar, traditional English units of measurement. The metric system has been used internationally in engineering and scientific fields for many years. It was not until the 1970s that there was an international movement for the global adoption of the metric system. Although the U.S. made an attempt in the mid-1970s to adopt the metric system, the adoption failed, and we remain one of the last holdouts. As students in environmental science, this dual system causes confusion and problems (e.g., the infamous case involving the Hubble Space Telescope). Nevertheless, to practice and understand science, you need to know the metric system. And to communicate your findings to the general public, you need to know how to convert to traditional English units of measurement. There are numerous on-line references for conversions (One example is the National Institute of standards and Technology: http://ts.nist.gov/ts/htdocs/200/202/mpo_home.htm; another is OnlineConversion at http://www.onlineconversion.com/). Most introductory science textbooks have conversions in their appendices. Below are some common metric units and conversions.

Length

1 centimeter = 10 millimeters
1 decimeter = 10 centimeters or 100 millimeters
1 meter = 10 decimeters or 100 centimeters or 1000 millimeters
1 kilometer = 1000 meters or 100,000 centimeters or 1,000,000 millimeters

Area

1 cm^2 = 100 mm^2
1 m^2 = 1,000,000 cm^2
1 hectare = 10,000 m^2
1 km^2 = 100 hectares or 1,000,000 m^2

Volume

1 centiliter = 10 ml
1 deciliter = 100 ml
1 liter = 10 dl or 1000 ml
1 m^3 = 1000 liters

Mass

1 kg = 1000 gm
1 tonne = 1000 kg or 1,000,000 gm

Conversion Factors

English to Metric Conversion Table

Change	From	To	Multiply by
LENGTH			
	feet	meters	0.3048
	inches	millimeters	25.4
	inches	centimeters	2.54
	miles	kilometers	1.6093
	miles	feet	5280
	yards	meters	0.9144
	yards	miles	0.0005682
VOLUME			
	cubic feet	cubic meters	0.0283
	cubic yards	cubic meters	0.7646
	gallons	liters	3.7853
	pints (dry)	liters	0.5506
	pints (liquid)	liters	0.4732
	quarts (dry)	liters	1.1012
	quarts (liquid)	liters	0.9463
WEIGHT			
	ounces	grams	28.3495
	pounds	kilograms	0.453592
	pounds	metric tons	0.000453592
	tons (short)*	metric tons	0.907185
AREA			
	acres	hectares	0.4047
	square feet	square meters	0.0929
	square miles	square kilometers	2.59
	square yards	square meters	0.8361

*Short ton = American ton (2,000 lbs).

Miscellaneous Conversion Table

Change	From	To	Multiply by
ENERGY			
	kilowatt-hour	BTU	3412
	watts	BTU/hour	3.4121
SPEED			
	feet/second	meters/second	.3048
	miles/hour	kilometers/hour	1.60934
TIME			
	hours	days	0.04167
	days	years	0.00273785
TEMPERATURE			
	degrees F	degrees C	$-32 \div 1.8$

Numerical Prefixes

The International System of Units uses the following prefixes to represent large decimal numbers. The following is a list of the more common units you will use in this lab:

Positive Numbers			Negative Numbers		
PREFIX	SYMBOL	NUMBER	PREFIX	SYMBOL	NUMBER
peta	P	10^{15}	milli	m	10^{-3}
tera	T	10^{12}	micro	μ	10^{-6}
giga	G	10^{9}	nano	n	10^{-9}
mega	M	10^{6}	pico	p	10^{-12}

About the Authors

Travis Wagner is an assistant professor in environmental science and policy in the Department of Environmental Science, University of Southern Maine. He received his Ph.D. in environmental and natural resource policy from The George Washington University, his M.P.P. in environmental policy from the University of Maryland-College Park, and his B.S. in environmental science from Unity College.

Robert Sanford is an associate professor of environmental science and policy in the Department of Environmental Science, University of Southern Maine. He received his Ph.D. and M.S. in environmental science from the State University of New York College of Environmental Science and Forestry and a B.A. in Anthropology from the State University of New York College at Potsdam.

Printed in the United States
203662BV00004B/1-18/A